DOUBLE-HULL TANKER LEGISLATION

AN ASSESSMENT OF THE OIL POLLUTION ACT OF 1990

Committee on Oil Pollution Act of 1990 (Section 4115)
Implementation Review

Marine Board
Commission on Engineering and Technical Systems
National Research Council

NATIONAL ACADEMY PRESS
Washington, D.C. 1998

NATIONAL ACADEMY PRESS • 2101 Constitution Avenue, N.W. • Washington, DC 20418

NOTICE: The project that is the subject of this report was approved by the Governing Board of the National Research Council, whose members are drawn from the councils of the National Academy of Sciences, the National Academy of Engineering, and the Institute of Medicine. The members of the panel responsible for the report were chosen for their special competencies and with regard for appropriate balance.

This report has been reviewed by a group other than the authors according to procedures approved by a Report Review Committee consisting of members of the National Academy of Sciences, the National Academy of Engineering, and the Institute of Medicine.

The project is part of a program supported by interagency agreement No. DTMA91-94-G00003, which is managed on behalf of participating agencies by the Maritime Administration of the Department of Transportation; and grant No. N00014-95-1-1205 between the Navy and the National Academy of Sciences. Financial support from the American Bureau of Shipping to cover the costs of report publication is gratefully acknowledged. The views expressed herein do not necessarily reflect the views of the sponsors and no official endorsement should be inferred.

Limited copies are available from: Marine Board, Commission on Engineering and Technical Systems, National Research Council, 2101 Constitution Avenue, Washington, D.C. 20418

International Standard Book Number 0-309-06370-1

Printed in the United States of America

COMMITTEE ON OIL POLLUTION ACT OF 1990 (SECTION 4115) IMPLEMENTATION REVIEW

DOUGLAS C. WOLCOTT *(chair)*, Chevron Shipping Company (retired), Ross, California
PETER BONTADELLI *(vice chair)*, California Department of Fish and Game, Sacramento
LARS CARLSSON, Concordia Maritime AB, Göteborg, Sweden
WILLIAM R. FINGER, ProxPro, Inc., Friendswood, Texas
RAN HETTENA, Maritime Overseas Corporation, New York, New York
JOHN W. HUTCHINSON, **NAE/NAS**, Harvard University, Cambridge, Massachusetts
SALLY ANN LENTZ, Ocean Advocates, Columbia, Maryland
DONALD LIU, American Bureau of Shipping, New York, New York
DIMITRI A. MANTHOS, Admanthos Shipping Agency, Inc., Stamford, Connecticut
HENRY MARCUS, Massachusetts Institute of Technology, Cambridge
KEITH MICHEL, Herbert Engineering Corporation, San Francisco, California
JOHN H. ROBINSON, Consultant, Santa Barbara, California
ANN ROTHE, Trustees for Alaska, Anchorage
DAVID G. ST. AMAND, Navigistics Consulting, Boxborough, Massachusetts
KIRSI K. TIKKA, Webb Institute, Glen Cove, New York

Marine Board Staff

CHARLES BOOKMAN, Director (until April 1997)
PETER JOHNSON, Acting Director (from May 1997)
DONALD PERKINS, Study Director (until June 1996), Consultant (from July 1996)
JILL WILSON, Study Director (from July 1996)
RICHARD WILLIS, Consultant
SHARON RUSSELL, Administrative Assistant (until June 1995)
MARVIN WEEKS, Administrative Assistant (from July 1995 to May 1997)
THERESA FISHER, Administrative Assistant (from May 1997)

Liaison Representatives

KEVIN BAETSEN, Military Sealift Command, U.S. Navy, Washington, D.C.
J. ROWLAND HUSS, Naval Sea Systems Command, U.S. Navy, Washington, D.C.
ZELVIN LEVINE, Office of Environmental Affairs, Maritime Administration, Washington, D.C.
FREDRICK SCHEER, Office of Standards Evaluation and Development, U.S. Coast Guard, Washington, D.C.
JAIDEEP SIRKAR, Office of Design and Engineering Standards, U.S. Coast Guard, Washington, D.C.

MARINE BOARD

Staff

Preface

Since the end of World War II, industrialized nations have imported increasing quantities of oil from the Middle East, the North Sea, Nigeria, Indonesia, the Caribbean, and other parts of the world. More than 3,300 tankers, each with a capacity of more than 10,000 deadweight tons (DWT), now serve the world maritime oil trade; approximately 40 percent of these vessels call each year at U.S. ports. Although the maritime oil trade supports economic growth in many countries, it has also raised concerns about damage to the marine environment in the event of oil spills.

As the demand for maritime oil transportation increased rapidly in the postwar years, the average size of a tanker grew. A single cargo tank on today's large tankers can hold more than twice as much oil as an entire World War II tanker. The large size of tank vessels and major spillage from vessel accidents—such as the grounding and breakup of the *Torrey Canyon* off the Scilly Isles in 1967—stimulated international action to formulate tank vessel design and construction standards aimed at reducing oil outflow following tanker damage. These standards, which are incorporated in international conventions, were developed by representatives of governments of the major international maritime nations and by industry representatives, ship classification societies,[1] and other interested parties, working under the auspices of the International Maritime Organization (IMO), a specialized agency of the United Nations. The IMO standards in MARPOL 73/78[2] addressed ballast tank location in tank vessel designs and

[1]For example, the American Bureau of Shipping, Lloyd's Registry of Shipping, and Det Norske Veritas.

[2]The International Convention for the Prevention of Pollution from Ships, 1973, as modified by the Protocol of 1978. This convention, known as MARPOL, addresses pollution from oil, chemicals, and other harmful substances, garbage, and sewage.

operational requirements such as ballast tank cleaning as a means of reducing oil outflow after ship collisions and during routine operations. Enforcement of these IMO standards was primarily dependent on the actions of flag states (nations where tank vessels are registered) and of classification societies. Since 1990, a review of procedures by both IMO and the classification societies has led to a strengthening of port-state enforcement options and increased stringency of internal classification society procedures aimed at increasing vessel quality.

The grounding of the *Exxon Valdez* in Prince William Sound in March 1989, and the subsequent spillage of more than 11 million gallons of crude oil into Alaskan waters, resulted in changes in both the character of tank vessel design standards and the manner in which they are formulated. In August 1990, the U.S. Congress promulgated P.L. 101-380, the Oil Pollution Act of 1990 (OPA 90). The intent of this law was, in part, to minimize oil spills through improved tanker design, operational changes, and greater preparedness. Section 4115 of OPA 90 focused on changes in ship design—notably double hulls—to prevent or minimize spillage when an accident occurs.[3] Single-hull tank vessels of 5,000 gross tons or more will be excluded from U.S. waters after 2010 unless they are equipped with a double bottom or double sides, in which case they may be permitted to trade to the United States through 2015, depending on their age. An exemption allows single-hull tankers trading to the United States to unload their cargo offshore at deepwater ports or in designated lightering areas through 2015.[4,5]

The fact that the United States, as a port state,[6] unilaterally promulgated legislation that applies to all tankers operating in U.S. waters, not just to U.S.-flag vessels, had a worldwide impact. Following the passage of OPA 90, changes in the international regulatory regime in the form of two additions to MARPOL 73/78 mandated a worldwide transition to double-hull vessels or their equivalents. MARPOL 73/78, Regulation I/13F (MARPOL 13F) specifies hull configuration requirements for new tankers of 600 DWT[7] capacity or greater contracted after July 1993; oil tankers of more than 5,000 DWT are required to have double hulls or the equivalent. MARPOL 73/78, Regulation I/13G (MARPOL 13G) addresses operational requirements to reduce oil outflow from single-hull vessels in the

---

[3]OPA 90, Section 4115 (c)(2) states that tank vessels shall be equipped with a double hull or "with a double containment system determined by the Secretary of Transportation to be as effective as a double hull for the prevention of a discharge of oil." The Secretary has not approved an equivalently effective system as of the date of this report.

[4]The Louisiana Offshore Oil Port is currently the only offshore deepwater port in the United States.

[5]In practice, very large crude carriers (VLCCs) are the primary users of lightering zones and the deepwater port, although the exemption applies to all tankers regardless of size.

[6]A port state is a nation whose ports are called on by any vessels of any flag.

[7]OPA 90, Section 4115 defines vessel sizes in gross tons (GT), whereas MARPOL 13F and 13G use DWT. GT is a volumetric measure of a vessel's size as determined according to international convention. DWT is a measure of the weight of cargo plus water, fuel, and stores that a vessel can carry.

world tanker fleet and specifies a schedule for retrofitting or retiring such vessels 25 or 30 years after delivery.

ORIGIN OF THE STUDY

Congress anticipated that OPA 90 would have significant and wide-ranging effects on both the domestic and the world tanker fleets; more than 90 percent of the tank ships calling on U.S. ports operate under a foreign flag. Congress ordered the U.S. Secretary of Transportation, acting through the U.S. Coast Guard (USCG), to assess the impact of Section 4115 on the marine environment and on the economic viability and operational makeup of the maritime oil transportation industry. After the USCG requested the advice of the National Research Council (NRC) in preparing its report to Congress, the NRC convened the Committee on the Oil Pollution Act of 1990 (Section 4115) Implementation Review under the auspices of the Marine Board. Committee members were selected with expertise in the following areas: tanker fleet management; tank vessel design, construction, operation, and maintenance; economics of oil sourcing and oil transportation; marine safety; marine environmental law and policy; natural resource damage assessment; international maritime conventions; and federal regulations related to marine petroleum transportation and operations. Biographical sketches of committee members are provided in Appendix A.

STUDY SCOPE AND CONTEXT

The committee was charged with assessing the impact of the double-hull and related provisions of OPA 90, Section 4115 (see Appendix B) on three areas identified in the legislation:

Ship Safety and Protection of the Marine Environment. Determine the extent to which there has been a change (or the extent to which change can be anticipated) in oil pollution in U.S. waters; in the incidence of marine casualties; in the risk of oil spills resulting from, or influenced by, early retirement of tank vessels and exemptions to OPA 90; and in measures taken to modify single-hull tank vessels to reduce risk of accidental spillage (in compliance with OPA 90). Document the progress made in double-hull tank vessel design, construction, maintenance, and operations, and specifically identify any known safety problems that have occurred with double-hull tank vessel designs.

Economic Viability of the Maritime Oil Transportation Industry. Determine the effect of OPA 90, Section 4115 on industry as may be evidenced, for example, by the extent of shifts to other modes and means of transportation, trends in shipbuilding and chartering, and changes in chartering rates. Identify added costs of construction and maintenance of double-hull tank vessels compared to non-double-hull tank vessels.

Operational Makeup of the Maritime Oil Transportation Industry. Identify the nature and extent of changes within the industry and the safety implications that may be related to OPA 90, Section 4115—for example, changes in tank vessel ownership and tank vessel type utilization.

In addition, the USCG and the NRC agreed that the scope of the assessment should include the influence of international conventions on tank vessel design and operation. In particular, the committee was asked to review and comment on evidence regarding the influence of international conventions (primarily MARPOL 13F and 13G) concerning hull design (for reducing the risk of oil spills from tank vessels) on the composition and character of tanker fleets and the interaction of these conventions with Section 4115.

OPA 90 addresses not only structural design issues, but also oil pollution liability and compensation, spill response planning, manning standards, vessel traffic services, and other issues. As a result, the maritime oil transportation industry has revised its operations, particularly in light of the law's strict liability provisions for oil spills and the potential costs associated with cleanup and related third-party and natural resource damage if spills occur. These changes come at a time when both the market for construction of new tankers and oil shipping rates (freight rates) are emerging from a depressed period, during which income was usually insufficient to cover the cost of new investment.

Changes in the international regulatory environment are also affecting tank vessels. In addition to the structural and operational requirements of MARPOL 13F and 13G, initiatives of note include enhanced surveys by classification societies, increased audits and inspections of vessels by charterers and the sharing of this information through industry-sponsored programs, and more comprehensive port-state control activities. These factors combine to affect the safety of the overall fleet by preventing casualties that could result in oil spills, reducing oil outflow from casualties, or decreasing the number of tank vessels subject to casualties. Because the influence of Section 4115 on safety is intertwined with other factors, its effects are difficult to isolate, and this complexity is reflected in the committee's findings.

The committee's assessment, moreover, was subject to constraints inherent in the timing of the study. Insufficient data on actual incidents were available to evaluate the effect of Section 4115 on oil spills from vessels in U.S. waters. The committee's assessment, therefore, is based on an analytical comparison of the oil outflow characteristics of double-hull and single-hull designs. In accordance with its charge, the committee did not question the double-hull mandate or examine alternative designs potentially equivalent to double hulls. A more comprehensive discussion of alternative tank vessel designs can be found in *Tanker Spills: Prevention by Design* (NRC, 1991).

With respect to tank barges, the committee's assessment focused on barges engaged in the ocean transportation of petroleum. The spill data presented in Chapter 2 include spills from both inland and oceangoing barges, but the economic and structural analyses in Chapters 5 and 6 are limited to oceangoing barges.

Study Methods

The full committee met six times over the course of the study. In addition, several committee members held work sessions to analyze specific topics and data and to draft sections of the report. Several supplementary studies were conducted under subcontract (see Appendix D), most notably a comparative analysis of double-hull and single-hull tank vessel designs performed by Herbert Engineering Corporation (see Appendix K).

To obtain the necessary data, the committee made an exhaustive search of available data resources in the public and private sectors concerning the following: double-hull construction and safety; early retirement of single-hull vessels; tanker fleet composition and ownership; international maritime rules; the lightering and deepwater port exemption to OPA 90; oil spills and oil spill risk; oil supply and demand; single-hull modification; tanker economics and operations; and vessel casualties. In addition, publications and reports related to the study topic were reviewed by the committee. Files developed by the USCG since the initiation of OPA 90 were a significant resource, as were unpublished maritime accident data for 1994 and 1995 obtained with the assistance of the USCG.

Industry representatives provided the committee with current information on a number of topics, including maritime oil industry economics, tanker sale and purchase brokerage, shipbuilding trends and costs, trends in inspection practices for double-hull tank vessels, vessel finance and insurance, and the operational and economic characteristics of the oceangoing domestic barge fleet. A list of presentations made to the committee is provided in Appendix D. The committee also sent questionnaires to shipyard operators and to the owners and operators of double-hull tankers, designers of double-hull tankers, classification societies, and oceangoing tank-barge operators to solicit information on design trends, costs, problems with double-hull vessels, and any special concerns and practices unique to double-hull design (see Appendix C).

The committee requested public comments on its interim report (NRC, 1996) by means of a USCG announcement in the Federal Register in April 1996 (Federal Register, 1996). Comments were received from the American Institute of Merchant Shipping (now the U.S. Chamber of Shipping), VELA International Marine Ltd. of Saudi Arabia, the State of Washington Office of Marine Safety, and the Water Quality Insurance Syndicate, an association of companies that insures vessel owners and operators against statutory and third-party pollution liability.

Organization of the Report

The report is divided into seven chapters and a series of appendixes. Chapter 1 provides an overview of oil demand and supply factors that determine the need for maritime oil transportation and describes the hull design characteristics of the world tanker fleet. Chapters 2 and 3, respectively, address the safety and protection of the marine environment and the operational makeup of the marine oil transportation industry. The economic impact of OPA 90, Section 4115 on the world tanker fleet and on the domestic fleet of tankers and oceangoing barges is addressed in Chapters 4 and 5, respectively. Chapter 6 discusses issues concerning tank vessel safety, construction, maintenance, and operations that have been matters of concern since the early debates that led to the promulgation of OPA 90. The conclusions and recommendations stemming from the findings of Chapters 2 through 6 are given in Chapter 7.

ACKNOWLEDGMENTS

The committee gratefully acknowledges the efforts of the many individuals and organizations who contributed their time and effort to this study in the form of presentations to the committee, correspondence, telephone calls, and responses to questionnaires and other requests for information. Particular thanks are given to Jaideep Sirkar of the USCG Office of Design and Engineering Standards: Jack Klingel of the USCG Office of Marine Safety, Security, and Environmental Protection; Zelvin Levine of the Maritime Administration Office of Environmental Activities; and Fred Scheer of the USCG Office of Standards Evaluation and Development.

In addition, the committee appreciates the assistance of individuals from various sectors of the marine oil transportation industry, including information and publication services, ship classification societies, and engineering organizations. Many of these individuals, identified in Appendix D, traveled from Europe or Asia to make presentations to the committee.

Finally, the chairman wishes to thank all the members of the committee for their hard work during meetings, for reviewing drafts of the report, and for their individual efforts in gathering information and writing sections of the report.

REFERENCES

Federal Register. 1996. Interim Report on Tank Vessel Design, Construction, and Operation Under the Oil Pollution Act of 1990. Notice of availability of interim report; request for public comments. FR 61(81):18457–18458. April 25.

National Research Council (NRC). 1991. Tanker Spills: Prevention by Design. Marine Board. Washington, D.C.: National Academy Press.

NRC. 1996. Effects of Double-Hull Requirements on Oil Spill Prevention: Interim Report. Marine Board. Washington, D.C.: National Academy Press.

Contents

The National Academy of Sciences is a private, nonprofit, self-perpetuating society of distinguished scholars engaged in scientific and engineering research, dedicated to the furtherance of science and technology and to their use for the general welfare. Upon the authority of the charter granted to it by the Congress in 1863, the Academy has a mandate that requires it to advise the federal government on scientific and technical matters. Dr. Bruce Alberts is president of the National Academy of Sciences.

The National Academy of Engineering was established in 1964, under the charter of the National Academy of Sciences, as a parallel organization of outstanding engineers. It is autonomous in its administration and in the selection of its members, sharing with the National Academy of Sciences the responsibility for advising the federal government. The National Academy of Engineering also sponsors engineering programs aimed at meeting national needs, encourages education and research, and recognizes the superior achievements of engineers. Dr. William A. Wulf is president of the National Academy of Engineering.

The Institute of Medicine was established in 1970 by the National Academy of Sciences to secure the services of eminent members of appropriate professions in the examination of policy matters pertaining to the health of the public. The Institute acts under the responsibility given to the National Academy of Sciences by its congressional charter to be an adviser to the federal government and, upon its own initiative, to identify issues of medical care, research, and education. Dr. Kenneth I. Shine is president of the Institute of Medicine.

The National Research Council was organized by the National Academy of Sciences in 1916 to associate the broad community of science and technology with the Academy's purposes of furthering knowledge and advising the federal government. Functioning in accordance with general policies determined by the Academy, the Council has become the principal operating agency of both the National Academy of Sciences and the National Academy of Engineering in providing services to the government, the public, and the scientific and engineering communities. The Council is administered jointly by both Academies and the Institute of Medicine. Dr. Bruce Alberts and Dr. William A. Wulf are chairman and vice chairman, respectively, of the National Research Council.

Tables, Figures, and Boxes

TABLES

FIGURES

BOXES

Executive Summary

The passage of the Oil Pollution Act of 1990 (OPA 90) by the U.S. Congress and subsequent modifications of international maritime regulations—namely, the addition of Regulations 13F and 13G to the International Convention for the Prevention of Pollution from Ships, adopted in 1973 and amended in 1978 (MARPOL 73/78)—resulted in a far-reaching change in the design of tank vessels: double-hull rather than single-hull tankers are now the industry standard. Section 4115 of OPA 90 excludes single-hull tank vessels of 5,000 gross tons or more from U.S. waters from 2010 onward, apart from those with a double bottom or double sides, which may be permitted to trade to the United States through 2015, depending on their age. Commencing in the year 2000, however, all Aframax and most Suezmax tankers[1] without double bottoms or double sides that exceed 23 years of age will be barred from U.S. trade. An exemption to OPA 90 allows single-hull vessels to use U.S. deepwater ports or lightering areas[2] until 2015. The international fleet governed by MARPOL is to be composed entirely of double-hull vessels (or approved alternatives) no later than 2023. Thus, nearly all vessels in the world maritime oil transportation fleet are expected to have double hulls by about 2020. The proportion of double-hull tankers in the world fleet increased from 4 percent in 1990 to 10 percent in 1994. This percentage is expected to grow rapidly between now and 2000 as new double-hull vessels replace many of the single-hull tankers constructed during the building boom of the mid-1970s.

---

[1]The size ranges for Aframax and Suezmax tankers are commonly defined as being 80,000 to 105,000 DWT (deadweight tons) and 120,000 to 165,000 DWT, respectively. The upper size limits, however, are sometimes quoted as 120,000 DWT for Aframax tankers and 200,000 DWT for Suezmax tankers.

[2]Lightering is the process of transferring cargo at sea from one vessel to another.

PROTECTION OF THE MARINE ENVIRONMENT

The promulgation of OPA 90 was in large part a response to public concern over the 1989 *Exxon Valdez* incident, in which more than 11 million gallons of crude oil were spilled into Alaskan waters. Such incidents involving spillage of more than a million gallons of oil have dominated spill statistics over the past two decades and have focused public attention around the world on the potential hazards of oil spills from large tankers. Compared to earlier five-year periods, there was a decline in the quantity of oil spilled from vessels in U.S. waters[3] in the period 1991 to 1995, as well as a reduction in the number of spills of more than 100 gallons. In particular, there were no oil spills of greater than a million gallons from tank vessels in U.S. waters. Between 1991 and 1995, tankers accounted for only about 10 percent of the total oil spilled from vessels in U.S. waters. In contrast, inland and oceangoing barges together accounted for approximately half the total spillage from vessels and were involved in the majority of oil spills in U.S. waters during this period.

The reduction in oil pollution in U.S. waters between 1991 and 1995 cannot be attributed to the requirements of Section 4115, notably the double-hull mandate and the operational and structural requirements aimed at reducing the outflow of oil following incidents that involve single-hull tank vessels. The first compulsory retirements of single-hull vessels did not occur until 1995, and the final rules on operational and structural requirements were not issued until July 1996 and January 1997, respectively. Thus, the timing of actions relating to Section 4115 precludes the possibility that they had a significant impact on oil spills in U.S. waters between 1991 and 1995. Nonetheless, the committee's analytical comparison of double-hull and single-hull designs indicates that properly designed double hulls are potentially more effective than single hulls in preventing and mitigating oil outflow after a vessel casualty. As discussed later, some double-hull vessels (mostly less than 160,000 DWT [deadweight tons]) currently operating—specifically those without longitudinal subdivision through the cargo tanks—will not provide the enhanced environmental protection in all accident scenarios that would be provided by properly designed double hulls.

In the view of the committee, the reduction in oil spillage in U.S. waters between 1991 and 1995 was the result of a number of actions that are in process or emerging, notably: an increased awareness among vessel owners and operators of the financial consequences of oil spills and a resulting increase in attention to policies and procedures aimed at eliminating vessel accidents; actions by port states to ensure the safety of vessels using their ports; increased efforts by ship classification societies to ensure that vessels under their classification meet or exceed existing requirements; improved audit and inspection programs by

[3]U.S. waters are defined as waters subject to the jurisdiction of the United States, including the Exclusive Economic Zone.

charterers and terminals; and the increased liability, financial responsibility, and other provisions of OPA 90. There is a general perception within the maritime oil transportation community that the quality of vessels trading to the United States has improved in recent years, although the data available to the committee were insufficient to demonstrate any such improvement.

DESIGN OF DOUBLE-HULL TANK VESSELS

On the basis of its analytical comparison of single-hull and double-hull designs using probabilistic outflow methodology, the committee concluded that in the event of an accident involving a collision or grounding, an effectively designed double-hull tanker will significantly reduce the expected outflow of oil compared to that from a single-hull vessel. Similar analytical results were obtained for oceangoing barges. The committee concluded that complete conversion of the maritime oil transportation fleet to double hulls will significantly improve protection of the marine environment.

Despite the potential advantages of double hulls, not all double-hull vessels designed or built since 1990 provide the environmental protection and safe operation that were anticipated when the double-hull mandate was adopted. Certain designs, most notably those with "single-tank-across" cargo tank arrangements,[4] may exhibit excessive oil outflow following an accident and encounter intact stability[5] problems during cargo transfer operations even though they are in full compliance with design regulations of the International Maritime Organization (IMO) and major classification societies as of July 1997. The committee's analysis indicated that double-hull tankers with single-tank-across cargo tank arrangements have approximately twice the projected average outflow of tankers with longitudinal subdivision through the cargo tanks and also perform less well in the case of extreme outflow than single-hull vessels of pre-MARPOL or MARPOL design. In addition, of the nineteen double-hull designs analyzed, four—all with single-tank-across cargo tank configurations—can potentially become unstable during load and discharge operations. Several incidents involving instability of double-hull tankers at terminals in the United States and overseas have been reported. Intact stability and outflow concerns are significant because more than half of the vessels of less than 160,000 DWT in the current fleet of double-hull tankers have single-tank-across cargo tank arrangements.

These potential problems demonstrate clearly that the national and international design guidelines originally developed for single hulls are not suitable for double-hull designs. The committee is concerned that the United States, having taken the lead in mandating double hulls for vessels operating in its waters, has not assumed a leadership role in developing the technical guidelines needed to

[4]Such designs do not have longitudinal bulkheads through the cargo tanks.

[5]Stability when no damage has occurred is known as intact stability.

properly implement the legislation. Given the large number of new double-hull tankers likely to enter service within the next few years as tankers constructed during the mid-1970s boom are retired, there is a need to implement additional design guidelines as soon as possible. The use of performance-based design criteria would take account of the variations in performance of different double-hull designs and provide flexibility in developing potentially superior designs.

IMO has acted recently to address intact stability issues for both new and existing double-hull vessels. MARPOL Draft Regulation I/25A(2) establishes a "design-only" requirement to ensure the intact stability of new vessels; operational measures or a combination of design and operational measures are not acceptable options. The Marine Environment Protection Committee of IMO will circulate the draft regulation with a view toward adoption in September 1997. If the draft regulation is adopted, enforcement is expected in February 1999. In addition, an IMO circular[6] provides guidance on the operational measures needed to ensure adequate intact stability for existing double-hull tankers.

Outflow regulations are currently under development at IMO. The committee considers new design guidelines for outflow essential to ensure that the potential for environmental protection afforded by double-hull designs is fully realized in all new vessels.

The tanker owners and operators surveyed by the committee reported significant differences between double-hull and single-hull tankers in terms of operational safety, inspection and maintenance, and cargo operations. Despite some concerns about access to and ventilation of ballast spaces and about intact stability, industry representatives generally believe that double-hull tankers can be operated safely, albeit with additional resources and more attention than are needed to operate single-hull tankers. In the view of the committee, mandatory operational measures are necessary to ensure the safe operation of existing double-hull tankers with single-tank-across cargo tank arrangements.

Recommendation. The U.S. Coast Guard (USCG) should expand and expedite research efforts and cost-benefit evaluations necessary to develop rules appropriate for the design of double-hull tankers and tank barges. The following are of particular importance:

- Probabilistic analysis of oil outflow should be made an integral part of the design and review process for new double-hull tank vessels. Design requirements should ensure that all new double-hull tankers offer environmental performance at least equivalent to that provided by the IMO reference double-hull designs.
- Design requirements should include an assessment of intact stability through-

[6]The circular, entitled "Guidance on Intact Stability of Existing Tankers During Liquid Transfer Operations," does not constitute a regulation.

out the range of loading and ballasting conditions to identify potentially unstable conditions. Following the lead taken by IMO and to provide consistency with anticipated international requirements, adequate intact stability should be achieved by design.

Design rules should be implemented as soon as possible—if necessary in interim form—to ensure that all new double-hull tank vessels entering service do not pose a safety risk because of poor intact stability characteristics and have adequate internal subdivision to take full advantage of the spill-mitigating capabilities of double hulls.

Recommendation. The USCG should develop and implement operational procedures for existing double-hull tanker designs subject to intact stability problems. Such procedures should ensure adequate stability at all times during cargo transfer operations and should include appropriate crew training. Consistency between procedures for vessels in U.S. waters and corresponding international procedures is highly desirable.

OPERATIONAL MAKEUP OF THE MARITIME OIL TRANSPORTATION INDUSTRY

Aside from an increase in the proportion of double-hull tankers in the world fleet between 1990 and 1994, the committee could not definitively attribute changes in the makeup of the maritime oil transportation fleet since 1990 to either OPA 90 or MARPOL 13F and 13G. Growth in the percentage of independent ownership in both the world and the U.S. trading fleets, primarily at the expense of oil company ownership, reflects a decision by some major oil companies to leave the tanker business, in large part to avoid high-liability exposure as well as for other economic reasons. The vessel size distribution of the fleet trading to the United States has changed because of an increase in short- and medium-haul oil imports from Latin America and the Caribbean, which are carried in vessels of 80,000 to 150,000 DWT, and a reduction in long-haul oil imports from the Middle East, which are carried in very large crude carriers (VLCCs) of 200,000 DWT or more. Changes in the age distribution of the fleet trading to the United States reflect both the aging of vessels built during the boom of the mid-1970s and the relatively large number of newly constructed VLCCs. Vessels between 20 and 24 years of age and those up to 4 years of age carried more tonnage in 1994 than they did in 1990.

OPA 90 and MARPOL 13F and 13G have not yet had a significant impact on the age of vessels trading to the United States. Before the implementation of OPA 90, few vessels over 25 years of age traded to the United States. This situation may change, however, as a result of the aging of the VLCC fleet, the deepwater port and lightering zone exemption of OPA 90, and actions by other nations (such as Japan and Korea) to prevent or discourage older vessels from calling at their ports. It is probable that under the OPA 90 exemption, large single-hull vessels up to

30 years of age will operate to the United States through 2015 (see below). Measures will be needed to ensure that such vessels are adequately maintained and that their operation does not pose an unacceptable risk to the marine environment.

Recommendation. The USCG should implement a vessel surveillance program to ensure that the physical condition, maintenance, and operating procedures of vessels that are permitted to discharge their cargo offshore, but are barred from shore ports by the phaseout provisions of Section 4115, are held to appropriate levels. For example, the frequency and standards of inspection defined in the Port State Inspection Program and applied to vessels using non-offshore ports might also be applied to vessels using lightering areas and the U.S. deepwater port.

ECONOMIC VIABILITY OF THE MARITIME OIL TRANSPORTATION INDUSTRY

International Tanker Fleet

The impact of the double-hull requirement on the international tanker industry will be driven by MARPOL 13F and 13G and by Section 4115 of OPA 90. Although the latter will gradually bar single-hull tankers from trading to the United States, it will not necessarily force them into retirement from non-U.S. trade. MARPOL 13G, on the other hand, mandates the retirement of all single-hull tankers in international trade at 30 years of age. To trade beyond 25 years of age, pre-MARPOL tankers must retrofit protectively located spaces or make use of hydrostatically balanced loading (HBL)[7] in selected cargo tanks.

If historical trends continue, many tankers in international trade are likely to be scrapped before their statutory (MARPOL) retirement dates. In other words, their life expectancy will not be affected by legislation requiring double hulls. However, the economic factors influencing tanker lifetime may change, in part because of the double-hull mandates of MARPOL and OPA 90. The capital cost of a double-hull tanker is estimated to be 9 to 17 percent higher than that of a corresponding single-hull tanker, and operating and maintenance costs run 5 to 13 percent higher.

In the light of these increased costs, some owners of single-hull VLCCs and other large tankers that can trade economically to the U.S. deepwater port and lightering areas are expected to adopt HBL as a means of extending the operating life of their vessels from 25 to 30 years. The combination of HBL and the deepwater port and lightering zone exemption has virtually nullified the OPA 90

[7]For HBL, the level of cargo (e.g., crude oil) is limited to ensure that the hydrostatic pressure at the tank (and ship) bottom is less than the external sea pressure. Thus, if the tank is breached, seawater will flow in rather than oil flowing out.

age requirement for large single-hull tankers (150,000 DWT and more) that use HBL and are suitable for unloading within U.S. lightering areas or at the deepwater port. Without the OPA 90 exemption, such vessels over 25 years of age would be excluded from U.S. waters after 2010. Without the option of HBL life extension from 25 to 30 years (permitted by MARPOL 13G but not by OPA 90), such vessels would be excluded from international trade and would not be able to take on cargo for delivery to the United States. Smaller single-hull tankers, particularly those for which unloading offshore is not economical, may be forced into scrapping before the end of their expected economic life. Single-hull tankers of between 60,000 and 150,000 DWT (without double bottoms or double sides) will be excluded from trade to the United States when they reach 23 or 25 years of age, in accordance with the phaseout schedule of Section 4115.

The committee estimates that the cost of replacing the current single-hull world trading fleet of about 3,000 tankers—aggregating 280 million DWT—with new double-hull vessels and operating them through one 20-year life cycle will be about $30 billion greater than building and operating an equivalent single-hull fleet. This additional cost equates to approximately 10 cents per barrel of oil transported or about one-tenth of the cost of transportation, which is itself about 5 to 10 percent of the delivered cost of oil. Although current shipyard capacity is more than adequate to meet the world demand for new double-hull tankers, existing freight rates are insufficient to meet the full operating and construction costs of such vessels. Thus, freight rates are expected to rise as the industry transitions to double hulls. Given higher freight rates, the financial community expects that sufficient capital will be available to fund the conversion.

U.S. Domestic (Jones Act) Tank Vessel Trade

The impact of the double-hull requirement on the domestic (Jones Act) fleet[8] is expected to be much greater than its impact on the international tanker fleet. One reason for this is that the construction costs of Jones Act vessels are significantly higher than those of vessels in the international fleet, regardless of whether a vessel has a single or a double hull. Unlike vessels in the international fleet, Jones Act vessels will generally reach their mandated retirement dates before reaching the end of their economic life. A second reason is that there is considerable uncertainty over future demand for vessels in both the Alaskan crude oil trade and the coastal products trade. A decline in demand may not provide a sustained freight level over the vessel's life sufficient to recover investment in double hulls. Hence, new construction or the conversion of single-hull vessels to

[8]Under the terms of the Jones Act, shipping between any two points in the United States, including the movement of Alaskan oil, is restricted to U.S. registered vessels owned by U.S. citizens, crewed by U.S. seafarers, built in the United States without construction differential subsidies, and operated without operating differential subsidies.

double hulls will be discouraged, even though adequate shipyard capacity is available for these purposes.

The economic burden on the Jones Act fleet of transitioning to double-hull vessels and the resulting impact on domestic waterborne transportation capability—including possible disruptions in the supply of crude oil and products—are in urgent need of further review. In particular, concerns over national defense and the ability to meet the energy needs of the Northeast under extraordinary circumstances, such as severe winter weather and pipeline or refinery disruption, have to be addressed. The effect of uncertainties about the future state of the Jones Act market regulations should be included in the assessment.

Recommendation. The policy issues associated with the potential loss of domestic waterborne transportation capability should be carefully examined within the context of the double-hull mandate of Section 4115 and the committee's finding that properly designed double-hull vessels—including barges—are expected to offer enhanced environmental protection compared to single-hull designs. This examination should be undertaken by an independent body and should address the perspectives of all stakeholders, including tank vessel owners and operators, the oil industry and oil consumers, environmentalists, and state and federal regulators. The study should be initiated as soon as possible to ensure that policy determinations are made prior to potential supply disruptions or inefficient economic decisions.

NEED FOR BETTER DATA

The committee's analysis of oil spills in U.S. waters was complicated by difficulties in obtaining complete and reliable data. The USCG oil spill database is not readily available, even to technically competent, bona fide organizations interested in assessing progress in reducing the occurrence and severity of oil spill incidents. Data are of variable quality from year to year, in part because of major shifts in data system structure and emphasis over time. In the judgment of the committee, improvements in the USCG database in terms of consistency, completeness, and accessibility would be beneficial not only in quantifying progress toward national environmental goals but also in developing future regulations and facilitating industry planning.

The committee's efforts to identify changes in the quality of vessels trading to the United States since the promulgation of OPA 90 were also hampered by data deficiencies, including limitations in the USCG port-state inspection database. Many of the data available are subjective in nature, and it was difficult to establish valid comparisons between data for different years because of a lack of consistent metrics. An absence of data on individual vessels and operators also hindered the committee's assessment.

Recommendation. The USCG should ensure that its oil spill database—including information on cause—is capable of facilitating the analysis of trends and the comparison of accidents involving oil spills. This would benefit the development of future regulations aimed at preventing oil spills and would facilitate industry planning.

Recommendation. The USCG should ensure that its port-state inspection database permits meaningful comparisons and analyses of current and future port-state activities, particularly in regard to identification and assessment of trends in the quality of the tank vessel fleet.

1

Introduction

During the past 25 years, the United States, in common with other nations, has become increasingly concerned about oil spills associated with waterborne transportation of oil. Legislative initiatives aimed at reducing oil spills have evolved slowly and have been punctuated by serious tank vessel accidents (Figure 1-1). For the most part, legislation prior to 1990 had been developed under the auspices of the International Maritime Organization (IMO).

The U.S. Oil Pollution Act of 1990 (OPA 90, P.L. 101-380), however, was a major departure from the international effort to address shortcomings in tank vessel design and operation. In promulgating a requirement to change from single-hull to double-hull designs, the United States acted unilaterally. Section 4115 of OPA 90 excludes single-hull tank vessels of 5,000 gross tons or more from U.S. waters after 2010.[1] Section 4115 also requires tankers and barges without double hulls operating in U.S. waters to comply with interim regulations defining structural and operational requirements aimed at providing "as substantial protection to the environment as is economically and technologically feasible." The international community, through IMO, subsequently endorsed the goals of OPA 90 by implementing amendments to The International Convention for the Prevention of Pollution from Ships, adopted in 1973 and modified in 1978 (MARPOL 73/78), requiring (1) double-hull vessels or their equivalents in virtually all the world's tanker trades and (2) additional operational and structural measures for single-hull tank vessels.[2]

[1]Single-hull vessels unloading oil at deepwater ports or off-loading in lightering zones may operate through 2015. Vessels with a double bottom or double sides may be permitted to trade to the United States until 2015, depending on their age.

[2]Regulations 13F and 13G, respectively, of Annex I of MARPOL 73/78.

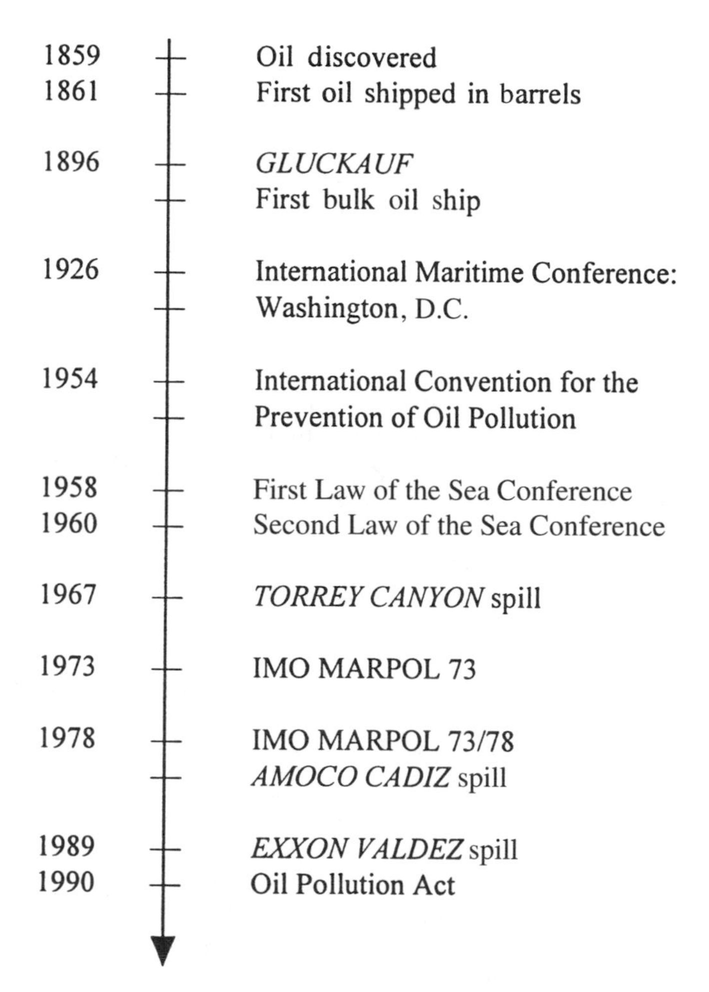

FIGURE 1-1 History of marine oil transportation and related legislation. Source: Noble, 1993. Reprinted with permission of the Society of Naval Architects and Marine Engineers (SNAME), 601 Pavonia Avenue, Jersey City, NJ 07306.

In enacting OPA 90, the U.S. Congress recognized that Section 4115 would have broad and potentially unexpected impacts. Congress therefore included language in the act that requires the Secretary of Transportation to review and assess its effects (see Appendix B). The purpose of the present study, requested by the U.S. Coast Guard, is to assist the Secretary of Transportation in assessing the measures mandated by Section 4115 and to ascertain their effects on protection of the marine environment, ship safety, the economics of oil transportation, and the makeup of the maritime oil transportation industry.

U.S. OIL SUPPLY AND DEMAND

Oil is supplied to the United States from both domestic production and foreign imports. The domestic supply comes primarily from crude oil production, supplemented by a small amount of natural gas liquids and a minor amount of other liquids. Foreign oil imports come in three forms: (1) crude oil, (2) unfinished oil materials ready for further processing or blending, and (3) finished petroleum products. The U.S. Department of Energy projects an increase in U.S. oil consumption of 3.5 million barrels per day (MBD) between 1994 and 2015 (EIA, 1996).

Between 1990 and 1994, domestic oil supply ranged from 9.6 to 10.1 MBD, but domestic oil production is expected to decrease by 1.2 MBD between 1994 and 2005, in part because of reduced output from the Alaskan North Slope (EIA, 1996). An increase in domestic supply of 0.9 MBD is anticipated between 2005 and 2015, however, as production from the lower 48 states increases.

Between 1990 and 1994, U.S. net oil imports increased from 7.2 to 8.1 MBD. As a result of increasing demand and decreasing domestic supply, net oil imports are projected to increase from 8.1 MBD in 1994 to 11.4 MBD in 2005. The projected increase in imports thereafter is modest, with imports in both 2010 and 2015 anticipated to be 11.8 MBD (EIA, 1996).

In 1994, U.S. consumption of oil products was about 17.7 MBD. After reducing oil imports by the amount of oil exports, 54 percent of U.S. consumption was supplied from domestic sources. The remainder of supply came from imports, most of which were waterborne, although some were supplied overland, primarily by pipeline from Canada. In 1994, Canada provided the United States with about 1.3 MBD of oil imports. This amount was slightly more than the 0.9 MBD of oil exports in 1994. In earlier years, imports from Canada were less than oil exports. Since 1973, it can be said that U.S. oil exports, on a volume basis, are generally offset by imports from Canada. Thus, a comparison between U.S. production and waterborne imports provides a reasonable basis for assessing the significance of waterborne imports to U.S. oil supply. Imports of crude oil by water grew from about one-quarter of U.S. domestic oil production in 1973 to almost equal domestic production in 1994 (Figure 1-2). This waterborne movement occurs almost entirely (more than 99 percent) in foreign-owned or foreign-registered tankers.

MARINE OIL TRANSPORTATION SYSTEM

International Oil Transport

The international marine oil transportation system has grown dramatically since World War II (Figure 1-3). More than 1.7 billion tons of oil are transported annually by ship from producing and refining countries to the consuming

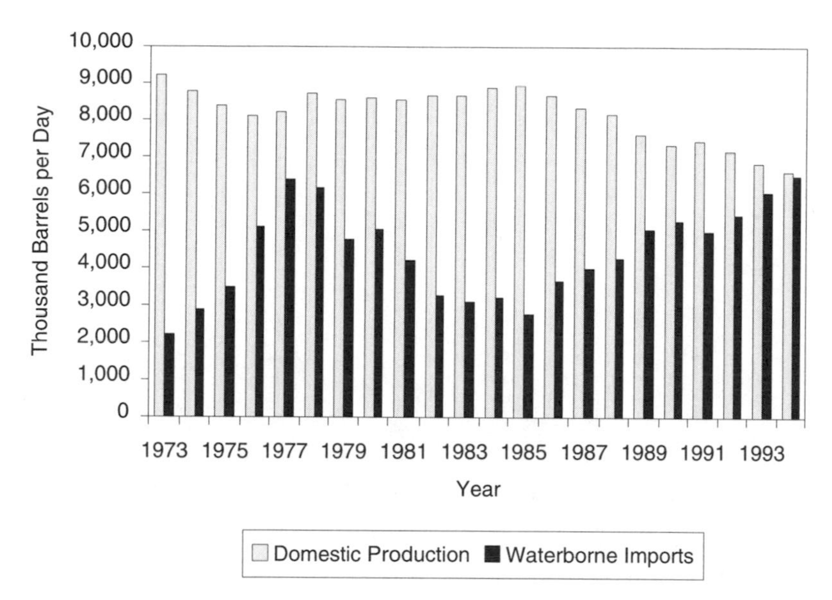

FIGURE 1-2 Waterborne crude oil imports and domestic crude oil production, 1973–1994. Sources: USACE, 1973–1994; EIA, 1995,

countries of the world. The United States presently accounts for about 30 percent of the total world waterborne oil movements. Thus, unilateral action by the United States in the area of tanker regulations can have profound effects internationally. In 1990, international oil movements (both crude oil and products) in U.S. waters totaled 513 million metric tons, and domestic coastal movements totaled 280 million metric tons (Lamb and Bovet, 1995).

As a result of the enormous increase in marine transportation of oil, the size of the largest tank vessels has increased from approximately 25,000 deadweight tons (DWT) at the end of World War II to more than 550,000 DWT today. The world fleet of tank vessels grew from 160 million DWT in 1971 to about 267 million DWT in 1994 (Drewry Shipping Consultants, 1994), having peaked at more than 300 million DWT in the late 1970s before economic conditions caused increased scrapping of vessels and a reduction in new construction (see Figure 1-4).

U.S. Oil Transport

The U.S. maritime transportation of oil and oil products has three distinct segments: the Pacific, Atlantic, and Gulf coasts. The Pacific coast trade is dominated by the Trans-Alaska Pipeline System (TAPS) crude oil trade, although there are increasing crude oil imports as well as a product fleet operating along the coast. The Atlantic coast trade involves significant movement of domestic oil

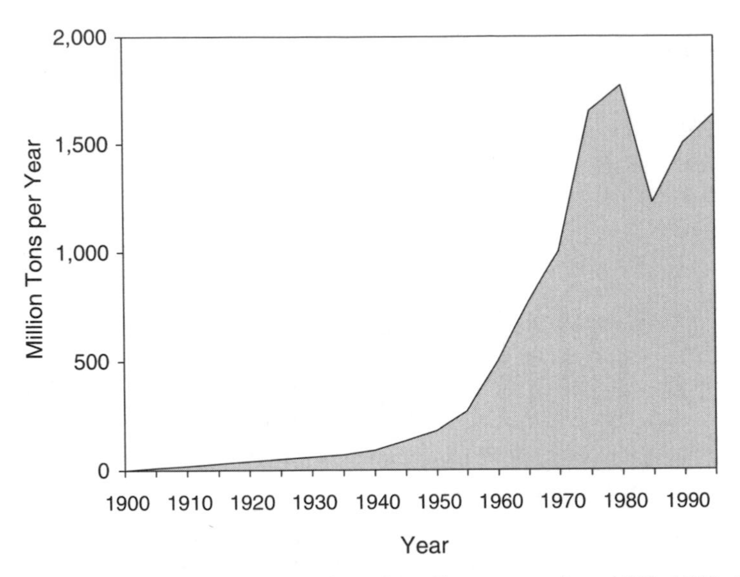

FIGURE 1-3 Growth in international marine oil transportation, 1900–1993. Source: Clarkson Research Studies Ltd., 1994. Reprinted with permission from Clarkson Research Studies, Ltd.

products from the Gulf coast and from mid-Atlantic refineries to the Northeast; imported products are also an important component of this trade. The substantial traffic in imported crude oil serves primarily the Gulf coast, Delaware River, and New Jersey refineries. In addition to the aforementioned tanker trades, barges operate in the coastal trades along the Gulf, roughly in parallel with coastal tankers but also serving smaller ports in intracoastal services.

Gulf coast traffic, which is greater than that of the other two coasts combined, accounts for 60 percent of all oil imported into the United States. Because larger tankers when fully loaded cannot enter the shallow waters of ports in the Gulf of Mexico to discharge their cargo, a two-phase system has developed to take advantage of the reduced carrying costs of the larger tankers. Vessels can discharge their cargo at the Louisiana Offshore Oil Port (LOOP), located 18 miles off the coast of Louisiana, or in lightering areas where smaller vessels take the oil directly to land-based terminals. About 30 percent of all imported crude oil delivered to the United States comes through the LOOP and lightering areas. The committee's analysis of Energy Information Administration projections (EIA, 1996) indicates that this could increase to about 50 percent over the next 10 years.

Economic Factors

The economics of transporting oil by water have changed dramatically over the years. In the early 1950s, marine transportation costs made up almost half the

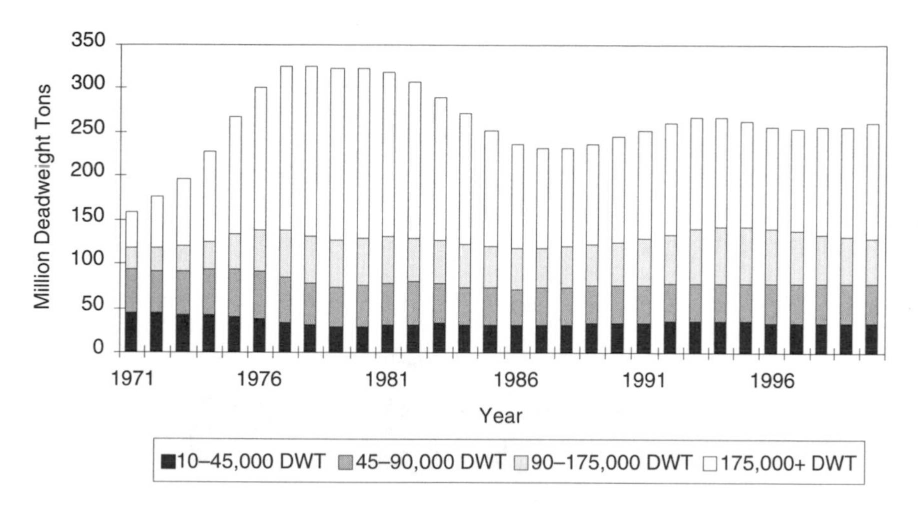

FIGURE 1-4 Oil tanker fleet development, 1971–2000. Source: Drewry Shipping Consultants, Ltd., 1994. Reprinted with permission from Drewry Shipping Consultants, Ltd.

price of oil delivered to a receiving dock. Since then, the price of oil at the producing location has increased tenfold, while the transportation cost has remained at approximately $1.00 per barrel in unadjusted dollars, or between about 5 and 10 percent of the delivered price. Part of the reason for this low transportation cost is the economy of scale achieved with larger tankers. In addition, a prolonged tanker surplus has led to depressed freight rates. Any increase in transportation costs caused by OPA 90 would be a very small portion of the total delivered cost of oil.

TANK VESSEL DESIGN

Prior to OPA 90, there were attempts to minimize oil spillage by means of vessel design, for example, by requiring double bottoms. In the late 1960s, the U.S. Coast Guard (USCG) and the administration initiated efforts to improve tanker design. The USCG led an effort at IMO to implement uniform international requirements to reduce pollution from tankers. This effort ultimately resulted in MARPOL 73, which promulgated the concept of segregated ballast to prevent operational discharges of oil residues in ballast water carried in cargo tanks. The 1978 Protocol modified the concept by requiring that for all new tank vessels of more than 20,000 DWT, segregated ballast tanks be located in a defensive way to reduce the chances that oil would be spilled in the event of grounding or collision. The actions required by Section 4115 aim to mitigate oil spillage by reducing the likelihood of a casualty occurring and reducing the amount of oil outflow given a casualty. The main purpose of the double hull is to reduce the probability of oil outflow following collision or grounding.

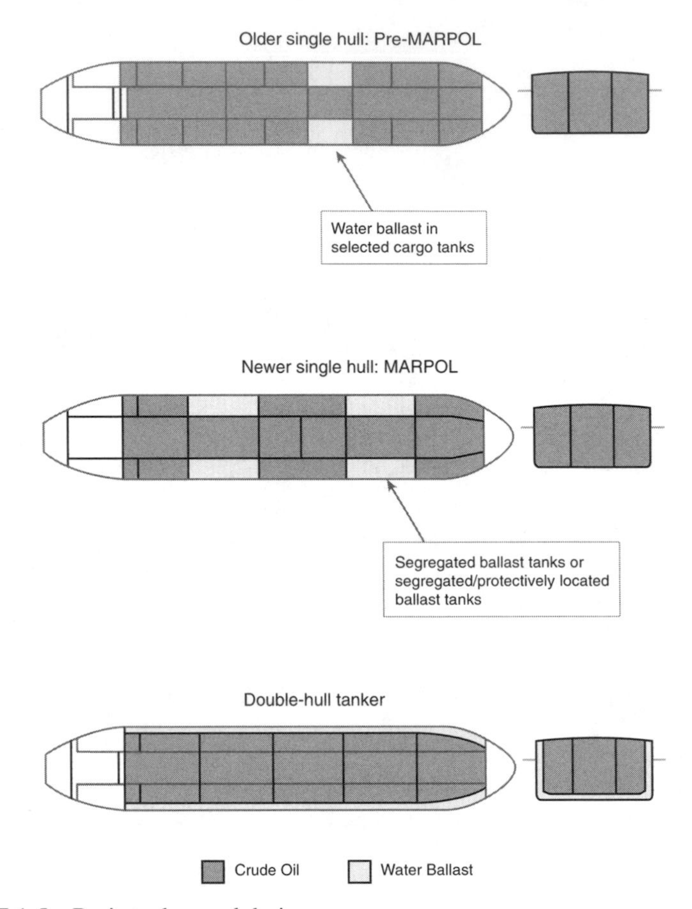

FIGURE 1-5 Basic tank vessel designs.

At the end of 1994, the world tanker fleet was made up of 1,732 pre-MARPOL tankers, 1,308 MARPOL tankers,[3] and 340 double-hull tankers (Tanker Advisory Center, Inc., 1995; see Figure 1-5). The percentage of double-hull tankers is expected to grow rapidly over the next few years as new double-hull vessels replace aging single-hull tankers constructed during the building boom of the mid-1970s.

---

[3]The committee has used the term "pre-MARPOL tanker" to describe vessels that carry ballast in cargo tanks and do not, for the most part, have segregated ballast tanks. The term "MARPOL tanker" is used to describe vessels that meet the segregated ballast requirements of MARPOL 73 or MARPOL 73/78.

REFERENCES

Clarkson Research Studies Ltd. 1994. Oil and Tanker Databook. London: Clarkson Research.

Drewry Shipping Consultants, Ltd. 1994. The International Oil Tanker Market: Supply, Demand, and Profitability to 2000. London: Drewry Shipping Consultants.

Energy Information Administration (EIA). 1995. Monthly Energy Review, February. Washington, D.C.: U.S. Department of Energy.

Energy Information Administration (EIA). 1996. Annual Energy Outlook 1996. Washington, D.C.: U.S. Department of Energy.

Lamb, B.L., and D.M. Bovet. 1995. An update on the Oil Pollution Act of 1990. Marine Technology 32(2):151–157.

Noble, P.A. 1993. Safer transport of oil at sea: A social responsibility for naval architects and marine engineers. Marine Technology 30(2):61–70.

Tanker Advisory Center, Inc. 1995. 1995 Guide to the Selection of Tankers. New York: Tanker Advisory Center.

U.S. Army Corps of Engineers (USACE). 1973–1994. Waterborne Commerce of the United States. Annual Review. Washington, D.C.: U.S. Army Corps of Engineers.

2

Ship Safety and Protection of the Marine Environment

The first part of the committee's task was to determine whether the marine environment has been better protected as a result of the Oil Pollution Act of 1990 (P.L. 101-380) (OPA 90) and whether any increased protection, if identified, could be linked to Section 4115. To carry out this task, the committee analyzed historical data on oil spills from vessels to determine (1) the effect of Section 4115 in reducing oil spills in U.S. waters; (2) the effect on oil spillage in U.S. waters of changes in the international regulatory regime since the enactment of OPA 90; and (3) the extent to which other government and industry initiatives have had an impact on oil spillage in U.S. waters.

In each case, difficulties were encountered in trying to attribute observed trends to legislative action or other initiatives. These difficulties derived in part from the timing of legislative and other actions, many of which are just now beginning to be implemented. Data limitations also contributed to problems in linking cause and effect. The committee therefore undertook a comparative analysis of single-hull and double-hull tanker and barge designs to determine the extent to which reductions in oil spills in U.S. waters can be expected as a result of the double-hull mandate of OPA 90. Some of the results of this analysis are summarized in the present chapter. More details are provided in Chapter 6.

HISTORY AND CAUSES OF SPILLS

The committee examined closely U.S. Coast Guard (USCG) records of oil spill incidents in U.S. waters during 1973–1995. These data are not readily accessible to the public or to research organizations interested in assessing U.S. progress in reducing oil spills. The committee, its contractors, USCG personnel, and others had to make extraordinary efforts to obtain adequate data for an assessment of oil spill trends, as required by the committee's charge. On close examination, it became clear that USCG data are not of uniform quality from year to year, and this problem was compounded by three major shifts in data system structure and emphasis over the years. Although many of the deficiencies pertained to the recording of smaller spills (less than 100 gallons), the committee also encountered problems in tracking larger spills. The USCG data were supplemented by information from the Minerals Management Service (MMS) of the U.S. Department of the Interior to assess several of the largest spills. The MMS database, although impressive in its overall quality, does not include information on spills of less than 1,000 barrels or spills from vessels other than tankers and barges. Thus, the MMS database is not a satisfactory substitute for a well-maintained and managed USCG database.

The committee's analysis revealed that spills from tank vessels (tankers and barges) have dominated the statistics over the years, accounting for 90 percent of the total volume of oil lost from all vessels since 1973 (see Figure 2-1).

Inspection of the tank vessel component of the spill data indicates that very large spills (those involving losses of more than 1 million gallons of oil) occurred with some regularity between 1973 and 1990 (see Figure 2-2). Out of a population of many thousands of smaller incidents, 18 spills, each of more than 1 million gallons, dominated the statistical pattern over the past two decades. One or two incidents could potentially have changed the overall statistical picture dramatically. However, there were no large spills involving losses of more than 1 million gallons during 1991–1995. USCG records dating back to 1973 indicate no similar period without a very large spill. Moreover, both the numbers of

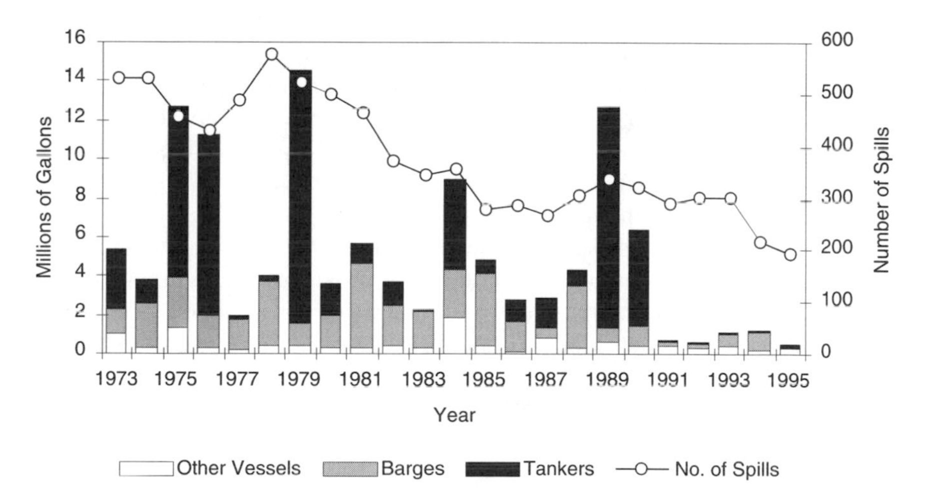

FIGURE 2-1 Number of oil spills and volume of spillage in U.S. waters, 1973–1995.

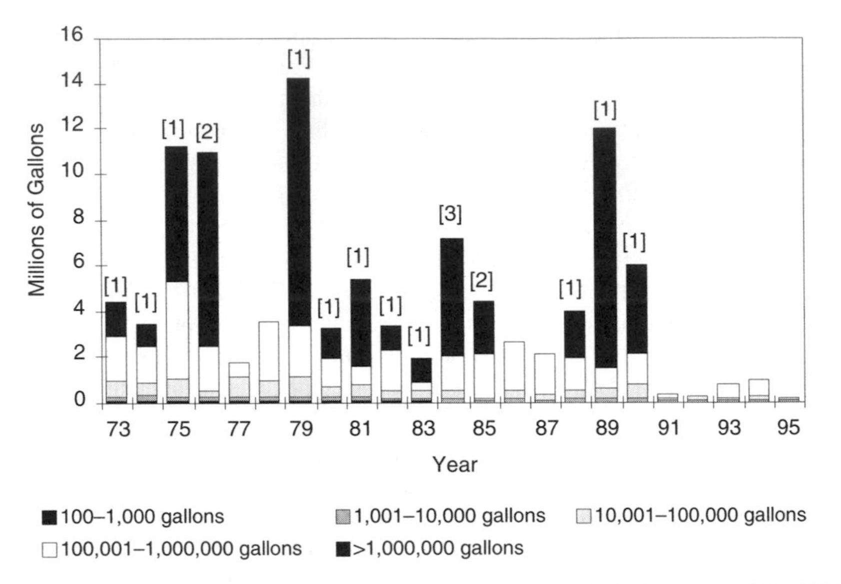

FIGURE 2-2 Volume of oil spilled from tankers and barges in U.S. waters, 1973–1995. Note: Numbers in brackets are spills of more than 1 million gallons.

incidents and the volumes released were at historically low levels during the period from 1991 to 1995.

Figure 2-3 summarizes the main types of casualty that resulted in oil spills from tankers and barges during the period 1991 to 1995, as indicated by USCG data. In the case of tankers (Figure 2-3a), the data show that a relatively small fraction of recent oil spillage was the result of collisions and groundings. The barge picture (Figure 2-3b) is more complex, with little consistency in the cause of spillage from year to year. The data presented in Figure 2-3 indicate that between 1991 and 1995, the total volume of oil spilled from barges in U.S. waters was significantly greater than that spilled from tankers. Tankers accounted for about 10 percent of the spillage from vessels in U.S. waters during this period, whereas barges accounted for approximately half of the total spillage from vessels and were involved in the majority of spills.

Given the OPA 90 phaseout schedule for single-hull vessels and delays encountered by the USCG in implementing other provisions of Section 4115 (see below), it is clear that the reasons for the improvement in the oil spill picture since 1990 cannot be attributed to the implementation of Section 4115.[1] The

[1]The USCG oil spill database includes the identity of the vessel for only about 10 percent of the recorded major casualties. Thus, the committee was unable to establish whether or not there is a correlation between vessel age and oil pollution. Such an analysis could have provided some indication of any changes in the risk of oil spills that might be anticipated as a result of the early retirement of tank vessels.

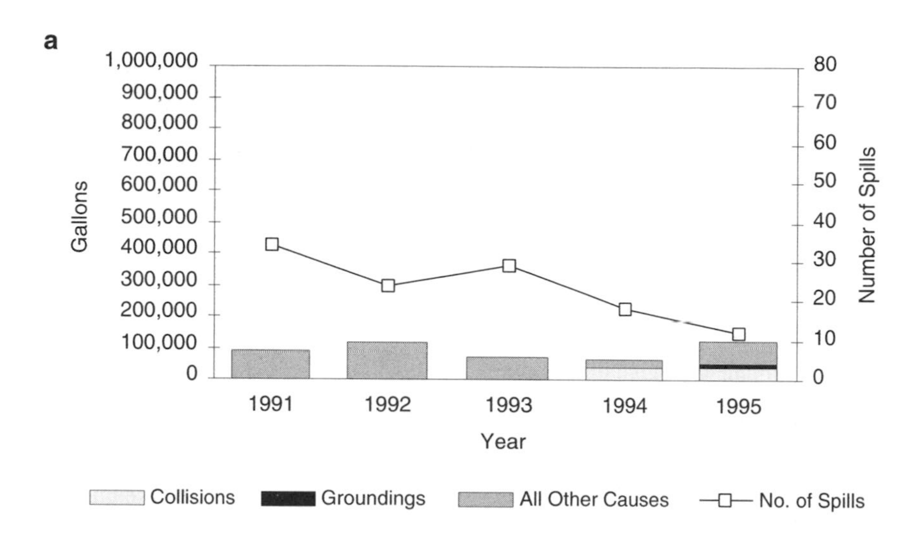

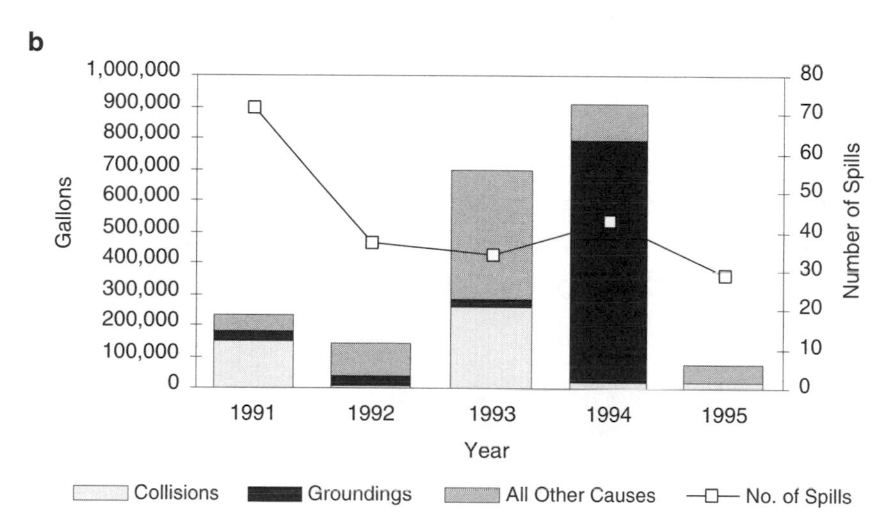

FIGURE 2-3 (a) Volume of oil spilled from tankers in U.S. waters and causes of spillage, 1991–1995. (b) Volume of oil spilled from barges in U.S. waters and causes of spillage, 1991–1995. Note: The data presented in Figure 2-3b include spills from both inland and oceangoing barges.

possible effects of international regulations and other government and industry initiatives are addressed below.

Despite the reduction in oil spilled over the past five years, the committee would caution against complacency. It is clear from the analysis presented above, as well as from practical experience with the *Exxon Valdez* incident, that the

TABLE 2-1 Section 4115 Phaseout Schedule for Vessels without Double Hulls by Age of Vessel

	Size of Vessel					
	5,000 to 14,999 GT		15,000 to 29,999 GT		30,000 GT or more	
Year of Double-Hull Compliance	Single Hull	Double Sides or Double Bottom	Single Hull	Double Sides or Double Bottom	Single Hull	Double Sides or Double Bottom
1995	40	45	40	45	28	33
1996	39	44	38	43	27	32
1997	38	43	36	41	26	31
1998	37	42	34	39	25	30
1999	36	41	32	37	24	29
2000	35	40	30	35	23	28
2001	35	40	29	34	23	28
2002	35	40	28	33	23	28
2003	35	40	27	32	23	28
2004	35	40	26	31	23	28
2005	25	30	25	30	23	28
2006	25	30	25	30	23	28
2007	25	30	25	30	23	28
2008	25	30	25	30	23	28
2009	25	30	25	30	23	28
2010	25	30	25	30	23	28
2011		30		30		28
2012		30		30		28
2013		30		30		28
2014		30		30		28
2015		30		30		28

Note: Vessels of ages shown, or older, must be phased out.

prevention of a single large spill can offer not only protection for the environment but also reduced costs for the vessel owner, the industry, and the nation as a whole. In addition to the vessel design issues identified in OPA 90 and addressed in this report, initiatives such as the USCG "Prevention Through People" program, which addresses the role of human factors in accident prevention, may further strengthen the ability to prevent the occasional, very large incident or to reduce its severity.

SECTION 4115 REQUIREMENTS AND IMPLEMENTATION

The principal requirements of Section 4115 that apply to vessels operating in U.S. waters are as follows:

- Single-hull vessels of 5,000 gross tons (GT) or more are excluded from U.S. waters beyond 2010. An exemption allows vessels without double hulls to operate to designated lightering areas or deepwater offshore oil ports until 2015.[2]
- Tankers with single hulls (or double bottoms or double sides) must be phased out according to a schedule that begins in 1995 and runs through 2015 (see Table 2-1).
- Existing tankers and barges without double hulls must comply with interim USCG regulations defining structural and operational requirements aimed at providing as substantial protection to the environment as is economically and technologically feasible.

Interim Structural and Operational Measures

The USCG has taken the following actions in developing and implementing the interim measures for single-hull vessels required by Section 4115.

A final rule defining requirements for emergency lightering equipment and the reporting of a vessel's International Maritime Organization (IMO) number prior to port entry was published on August 5, 1994 (Federal Register, 1994). This regulation is aimed primarily at ensuring that vessels carry minimum levels of lightering and deck cleanup equipment; it addresses after-the-fact spill actions rather than prevention or reduced outflows. A USCG final rule mandating operational measures to reduce oil spills from existing tank vessels without double hulls took effect on November 27, 1996 (Federal Register, 1996).[3]

---

[2]The relative operational and environmental risks associated with direct tanker deliveries of water-borne crude oil, offshore lightering and discharging at deepwater ports were assessed in a USCG study (1993).

[3]A final rule revising the underkeel clearance requirement for single-hull tank vessels and responding to petitions for rule-making will take effect on January 21, 1998 (Federal Register, 1997b).

A supplemental notice of proposed rule making (SNPRM) outlining several structural and operating alternatives under consideration by the USCG, such as hydrostatically balanced loading (HBL) and protectively located (PL) spaces,[4] was issued in December 1995 (Federal Register, 1995). The USCG final rule on structural measures, issued in January 1997, does not require vessels without double hulls to undertake any new structural measures in the remaining years before they are phased out (Federal Register, 1997a). This rule reflects the USCG finding that no structural measures can be retrofitted on existing single-hull tank vessels in a manner that is both technologically and economically feasible. The committee did not assess the validity of the USCG finding.

Only the regulation defining requirements for emergency lightering equipment and the reporting of a vessel's IMO number has been fully implemented to date.

Projected Reduction in Outflow with Double Hulls

The relative effectiveness of vessels with single and double hulls in mitigating oil outflow was assessed through probabilistic oil outflow analysis. Details of the analysis are provided in Chapter 6 and Appendix K. Thirty-six tankers and oceangoing barges, both single hull and double hull, were evaluated. The resulting probabilities of zero outflow for tankers are shown in Figure 2-4. The probability of zero outflow is defined as the likelihood that a vessel will be involved in a collision or grounding breaching the outer hull and not spill any oil. The reader is referred to Chapter 6 for a discussion of other important spill characteristics, notably mean and extreme outflow.

Over the past 15 years, collisions and groundings have been responsible for approximately 70 percent of the oil spillage from tankers and tank barges. In assuming that the current fleet is composed predominantly of single-hull vessels that are all replaced by double-hull vessels, the following changes are projected:

- Four out of every five oil spills attributable to collisions and groundings would be eliminated.
- A two-thirds reduction would be realized in the total volume of oil spilled from collisions and groundings.

These estimates are based on theoretical analysis. Future oil spill statistics will depend on many factors, including the extent of cargo tank subdivision incorporated into designs of double-hull tankers and tank barges. However, the relative numbers obtained from the analysis imply that the double-hull mandate, when fully implemented, will have a significant and positive effect on reducing the risk and the severity of oil spills.

---

[4]HBL and PL spaces are presently specified as interim alternatives for double hulls in the International Convention for the Prevention of Pollution from Ships, adopted in 1973 and amended in 1978, Regulation 13G of Annex I of MARPOL 73/78.

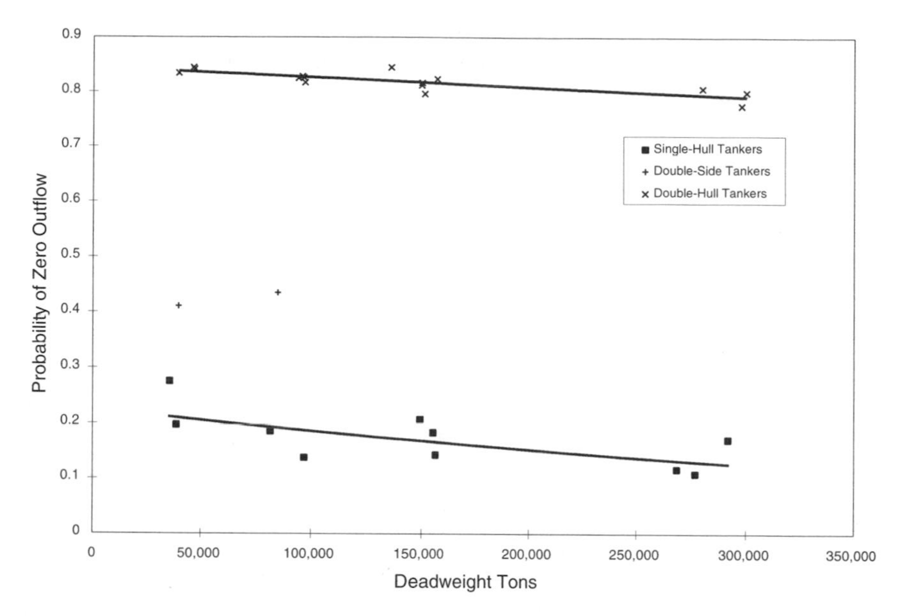

FIGURE 2-4 Probability of zero outflow for single-hull and double-hull tankers. Source: Herbert Engineering Corporation, 1996.

INTERNATIONAL REGULATORY REGIME

Significant events and actions have taken place in the international arena since the grounding of the *Exxon Valdez* and the passage of OPA 90 that may complement or enhance the actions required under OPA 90. Changes have occurred both in the international regulatory regime and as a result of port-state and industry initiatives. The former are discussed below; a later section of this chapter addresses government and industry initiatives.

Marine Pollution Prevention Prior to OPA 90

The International Maritime Organization, an agency of the United Nations, regulates international shipping through the adoption of international conventions by its members. Conventions that regulate ship design for safety and pollution prevention include the 1996 International Convention on Load Lines (ICLL), the 1974 International Convention on Safety of Life at Sea (SOLAS), and the 1973 International Convention for the Prevention of Pollution from Ships as amended by the Protocol of 1978 (MARPOL 73/78).

ICLL established the deepest draft to which a ship could be loaded to ensure that ships were not so overloaded as to run the risk of sinking or creating unsafe working conditions. SOLAS addressed the safety of the crew, passengers, ship, cargo, ports, and indirectly, the environment. Its principal provisions include general construction principles and requirements for ship subdivision and stability, safety equipment, fire protection, and radio telegraphy.

MARPOL 73/78 differs from ICLL and SOLAS in that it addresses the prevention of pollution from ships directly rather than indirectly. The 1973 convention required ballast to be carried only in clean or segregated ballast tanks (SBT),[5] thereby avoiding the pollution that can occur when ballast water containing remnants of oil is discharged from cargo tanks or when tanks are cleaned. The convention was amended in 1978 to require that segregated ballast tanks be located so as to provide protection against collisions and groundings (protectively located segregated ballast tanks [PL/SBT]).

Changes in Marine Pollution Prevention Following OPA 90

In November 1990, the United States submitted a proposal to the thirtieth session of the IMO Marine Environment Protection Committee to establish an international requirement for double-hull tankers. This proposal eventually resulted in the adoption of MARPOL 73/78 Regulations I/13F and I/13G (MARPOL 13F and 13G). These regulations, which became effective in July 1993, apply to the vessels of all nations and are similar to the provisions of Section 4115 of OPA 90.

New Vessel Requirements

MARPOL 13F specifies the hull configuration requirements for new oil tankers contracted on or after July 6, 1993, of 600 deadweight ton (DWT) capacity or more. Oil tankers between 600 and 5,000 DWT must be fitted with double bottoms or double sides, and the capacity of each cargo tank is specifically restricted. Every oil tanker of more than 5,000 DWT is required to have a double hull (double bottom and double sides), a mid-deck with double sides, or an alternative arrangement specifically approved by IMO as being equivalent to the double-hull design. These requirements, along with those of OPA 90, are shown in Table 2-2.

MARPOL 13F specifies that other designs may be accepted as alternatives to double hulls, provided they give at least the same level of protection against the release of oil in the event of collision or grounding and are approved, in principle, by IMO's Marine Environment Protection Committee. IMO design guidelines employ a probabilistic outflow methodology for calculating oil outflow and a pollution prevention index to assess the equivalency of alternative designs (see Chapter 6).

---

[5]Segregated ballast tanks are tanks designed for ballast only.

TABLE 2-2 Requirements of OPA 90 and IMO Regulation 13F for New Vessels

	Size	Hull Requirements	Enforcement Date
OPA 90 Section 4115	< 5,000 GT	Double-hull or double-containment systems[a]	Building contract placed after June 30, 1990 Delivered after January 1, 1994
	> 5,000 GT	Double hull	Building contract placed after June 30, 1990 Delivered after January 1, 1994
IMO Regulation 13F	< 600 DWT	Not applicable	
	600–5,000 DWT	Double bottom or double sides	Building contract placed after July 6, 1993 New construction or major renovation begun on or after January 6, 1994 Delivered after July 6, 1996
	> 5,000 DWT	Double hull, mid-deck with double sides, or equivalent	Building contract placed after July 6, 1993 New construction or major renovation begun on or after January 6, 1994 Delivered after July 6, 1996

[a]The double-containment system must be determined by the Secretary of Transportation to be as effective as a double hull in preventing a discharge of oil. As of this date, no double-containment system has been approved by the Secretary.

Existing Vessel Requirements

MARPOL 13G, which pertains to single-hull vessels, applies to crude oil tankers of 20,000 DWT or more, and to oil product carriers of 30,000 DWT or more. The regulation specifies a schedule for retrofitting (with double hulls or equivalent hull designs or operational measures) or retiring single-hull tank vessels 25 or 30 years after delivery. The differences between MARPOL 13G and OPA 90 are shown in Table 2-3.

Tankers not fitted with SBTs, or fitted with SBTs that are not protectively located, must designate protectively located double-side or double-bottom tanks or spaces when they reach 25 years of age. In appropriate locations, SBTs would be acceptable as protectively located spaces.

MARPOL 13G also permits HBL and other alternatives (operational or structural) to protectively located spaces. Tankers built in compliance with Regulation I (6) of MARPOL 73/78 (hereinafter referred to as MARPOL tankers) have protectively located ballast spaces and require no modification until

TABLE 2-3 Requirements of OPA 90 and MARPOL 13G for Existing Vessels

	Size	Hull Requirements	Enforcement Date
OPA 90 Section 4115	< 5,000 GT	Double-hull or double-containment systems	After January 1, 2015
	> 5,000 GT	Double hull	Per schedule starting in 1995
		Operational measures	November 27, 1996
MARPOL 13G	Crude carriers > 20,000 DWT and product carriers > 30,000 DWT	Double hull or equivalent	30 years after date of delivery
		PL/DS or PL/DB or PL/SBT or HBL or equivalent	25 years after date of delivery

Note: PL/DS = protectively located tanks, double sides; PL/DB = protectively located tanks, double bottom; PL/SBT = protectively located tanks, segregated ballast tanks; HBL = hydrostatically balanced loading.

reaching 30 years of age. On reaching 30 years of age, all tankers in the oil trade must be converted to double hulls or an acceptable equivalent according to MARPOL 73/78, Regulation I/13F(5).

The United States has reserved its position on the loading and structural provisions of MARPOL 13G applicable to single-hull tank vessels. The recent rule promulgated by the USCG does not require structural modifications of single-hull vessels before they are phased out. MARPOL 13G also imposes a program of enhanced ship inspections during periodic, intermediate, and annual surveys. This same provision is included in the November 1996 USCG rule on operational measures (Federal Register, 1996).

The fact that the United States has reserved its position on the aforementioned provisions of 13G will have little effect on most vessels calling at U.S. ports and on the resulting protection of U.S. waters. OPA 90 requires most vessels to retire by age 25, and 13G comes into effect only when vessels reach 25 years of age. Thus, most vessels 25 or older—whether in international or coastwise trade—will be excluded from U.S. waters by OPA 90, regardless of the provisions of 13G. There is one notable exception to this situation, namely, larger vessels operating to lightering areas and the deepwater port under the OPA 90 exemption. Tankers up to 30 years of age that are in compliance with 13G will be allowed to trade in international waters. These same vessels will be allowed to trade to the United States under the OPA 90 exemption, regardless of the U.S. position on 13G. The committee's concerns about the combined effects of 13G and the OPA 90 exemption, which together have virtually nullified the OPA 90 age requirement for large single-hull tankers, are addressed in Chapter 7.

Comparison of International Regulations with OPA 90, Section 4115

New Vessels

Section 4115 of OPA 90 and MARPOL 13F take different paths in addressing the change to double-hull construction. Section 4115 restricts oil trade to the United States by vessels without double hulls according to a schedule based on vessel age. MARPOL 13F takes a proactive approach by requiring all vessels constructed after a certain date to have double hulls or an approved alternative. MARPOL 13G allows existing vessels to trade for a longer period than that allowed under Section 4115 if they are of acceptable design. Figure 2-5 shows that Section 4115 is more restrictive in controlling vessels in the international fleet able to serve the United States.

Figure 2-6 shows the anticipated growth in petroleum tonnage carried in U.S. waters in double-hull vessels between 1990 and 2015 as a result of OPA 90 and

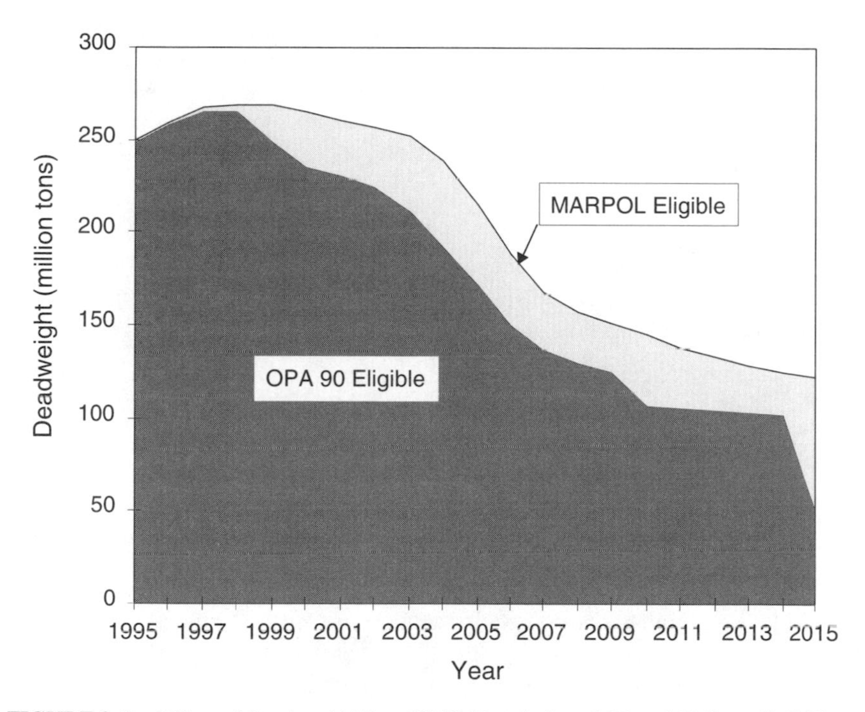

FIGURE 2-5 Effect of Section 4115 and IMO Regulations 13F and 13G on eligibility of existing vessels to operate in U.S. waters.

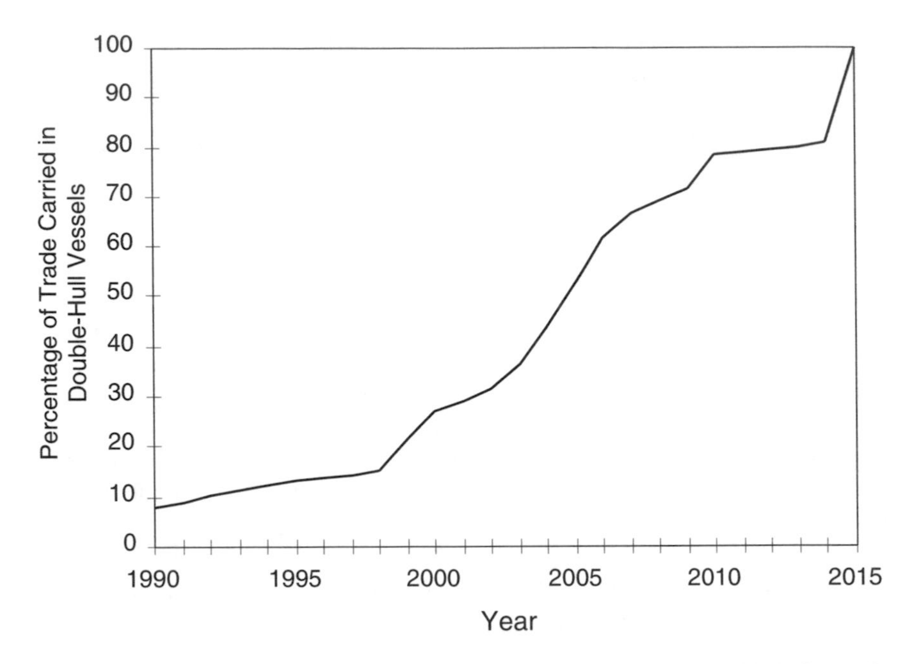

FIGURE 2-6 Increase in petroleum tonnage in U.S. waters carried in double-hull vessels. Note: The data used to generate this figure were obtained by the committee in the course of its study. Additional analysis was performed by individual committee members in the context of an ongoing study for the Volpe National Transportation Systems Center entitled *Regulatory Impact Analysis of OPA 90.*

MARPOL rules. The projection includes vessels in the international fleet trading to the United States and vessels in the domestic (Jones Act) fleet involved in coastal trade. Vessels without double hulls are assumed to trade until they reach 30 years of age (with an 8 percent loss of cargo capacity for those using HBL) unless they are excluded from U.S. waters by OPA 90. That is, the projection assumes that the deepwater port and lightering zone exemption to OPA 90 will extend the trading life of vessels without double hulls coming to U.S. waters to January 1, 2015.

Existing Vessels

Table 2-4 summarizes the requirements of the OPA 90 interim regulations and the international regulations for existing single-hull vessels. As with Section 4115, changes in vessel safety and the protection of the marine environment could not be attributed to the international measures because of their newness.

TABLE 2-4 OPA 90 and International Regulations for Tank Vessels without Double Hulls

	OPA, Section 4115(b)	International
Operational Measures		
Emergency lightering equipment	Prohibits use of cast iron or malleable iron valves or fittings	Flange specifications are IMO Universal Oil Transfer Connection specifications (33 CFR 154.500)
Bridge resource management policy and procedures	Written guidance to masters and officers in charge of navigational watch concerning need for continuously reassessing use of bridge-watch resources in accordance with bridge resource management principles	Consistent with Convention for Standards for Training, Certification, and Watchkeeping (STCW) Code Section B, Chapter VIII, Part 3-1
Vessel-specific watch policy and procedures	Written guidance to masters setting forth company policies and procedures to be followed to ensure all newly employed crew members are given a reasonable opportunity to become familiar with proper performance of their duties; on-the-job watch training for watchstanding personnel who assist the officer in charge of navigational watch	Consistent with STCW Code Section A, Chapter I, Part 14
Enhanced survey requirements	Dry-dock survey includes gauging report, visual inspection, and structure analysis; barge companies and smaller tankers allowed to have an alternative survey program with outside oversight or auditing	Incorporates IMO Resolution A.744(1E) by reference; IMO standard applies to > 20,000 DWT crude carriers and > 30,000 DWT product tankers; consistent with MARPOL 13G
Vital systems survey	Survey of mooring and cargo systems done by senior personnel	Incorporates International Safety Guide for Oil Tankers and Terminals (ISGOTT) guidance by reference; consistent with STCW Code Section A, Chapter V, Part 1
Autopilot alarm or indicator	Automatic alarm on autopilot required on tank ships; indicator on tugs	None

continued

TABLE 2-4 *Continued*

	OPA, Section 4115(b)	International
Maneuvering performance capability	Existing tank ships must calculate and post maneuvering performance	Incorporates IMO Resolution A.751(18) by reference; IMO resolution recommends that new vessels 100 meters or more in length meet performance standards; OPA 90 applies only to tank vessels > 5,000 GT
Maneuvering and vessel status information	Provide pilot cards and a wheelhouse poster	Incorporates IMO Resolution A.601(15) by reference
Minimum underkeel clearance	Plan, calculate, and log vessel's anticipated underkeel clearance prior to entering or departing port; liability cap is waived if done improperly	Ability of master and officers in charge to calculate ship's underkeel clearance inferred in STCW Section A, Chapter II, Part 1, but no specific underkeel clearance minimum required internationally
Emergency steering capability	Secondary steering system on primary towing vessels	None
Fendering system	Primary towing vessels and any assisting towing vessels must have enough fendering to prevent metal to metal contact between tug and barge	None
Structural Measures		
Protectively located segregated ballast tanks (PL/SBT)	None	Required for all non-SBT tankers using HBL and all non-PL/SBT tankers under MARPOL 13G
Alternatives to PL/SBT including HBL	None	Alternatives providing protection equivalent to PL/SBT subject to approval by member states under MARPOL 13G

GOVERNMENT AND INDUSTRY INITIATIVES

The recent U.S. and international regulations governing structural and operational practices in marine oil transportation are expected to have a significant long-term impact on oil spill prevention and the enhancement of ship safety. Port states and private organizations, including classification societies and industry groups, have developed programs unilaterally that have had a more immediate impact and are possibly the major reasons for the changes in spill patterns witnessed since the passage of OPA 90. Major initiatives that have been undertaken or improved since the adoption of OPA 90 include port-state quality controls, vetting by charterers,[6] and quality assurance programs.

Actions by Port States

Historically, port states (nations that have vessels calling at their ports) have not exercised substantial control over vessels that use their ports. Responsibility for vessel quality has been left to vessel owners, ship classification societies, and flag states (nations in which vessels are registered). However, in response to concerns that substandard vessels are still operating and increasing numbers of owners are registering their vessels in nations that do not meet their flag-state obligations, individual port states have intensified their control over shipping in recent years, aided by IMO and regional organizations. In 1994, more than 40,000 port-state control inspections were conducted in Europe, Canada, Australia, South America, and the United States.

Currently, four regional memoranda of understanding (MOU) have been signed to coordinate port-state programs: the Paris MOU (1982), the Acuerdo de Viña del Mar (1992), the Tokyo MOU (1993), and the Caribbean MOU (see Table 2-5).

IMO recently adopted port-state control guidelines that include criteria for qualifications and a code of conduct for port-state control officers.[7] Compliance with the International Safety Management Code (ISM) and revisions to the Convention for Standards for Training, Certification, and Watchkeeping (STCW) will be incorporated into port-state control programs.

In the United States, the USCG has adopted a more aggressive posture as a port state. Like Australia, which published its report the *Ships of Shame* (Parliament, 1992), the United States is targeting vessels for additional port-state surveillance and is making the resulting information available to the public (see Box 2-1). A key feature of the regional MOUs and the U.S. and Australian programs is the sharing of information among port states.

In an attempt to evaluate information on the condition of vessels calling on U.S. ports, the committee investigated the availability of USCG data from the ports of Houston, Los Angeles-Long Beach, and New York, describing the effectiveness of the ship surveillance program. Data collected in 1990 did not include information on the total number of vessels inspected or on vessel type. Therefore, it was not possible to make meaningful comparisons with 1995 data, and the

---

[6]Vetting is the quality assessment review of a particular vessel and its owner conducted by a charterer prior to entering into a chartering agreement.

[7]IMO Resolution A. 787 (19), Procedures for Port State Control, adopted at the nineteenth session of the assembly, November 23, 1995.

TABLE 2-5 Major Features of Regional Port-State Control Agreements

Agreement	Paris MOU	Acuerdo de Viña del Mar	Tokyo MOU	Caribbean MOU
Authorities that adhere to the MOU	Belgium, Canada, Denmark, Finland, France, Greece, Ireland, Italy, the Netherlands, Norway, Poland, Portugal, Spain, Sweden, the United Kingdom	Argentina, Brazil, Chile, Cuba, Uruguay	Australia, Canada, China, Hong Kong, Japan, Korea, Malaysia, New Zealand, Papua New Guinea, Russian Federation, Singapore, Vanuatu	Antigua and Barbuda, Aruba, the Bahamas, Barbados, Cayman Islands, Grenada, Jamaica, Netherland Antilles, St. Kitts and Nevis, Trinidad and Tobago
Authorities that have signed but have not yet accepted the agreement		Colombia, Ecuador, Mexico, Panama, Peru, Venezuela	Fiji, Philippines, Solomon Islands, Thailand, Vietnam	
Cooperating authorities	Croatia, Japan, Russian Federation, United States			
Observer authority	United States		United States	Anguilla, Canada, Montserrat, The Netherlands, United States, Caribbean Community Secretariat, Viña del Mar Secretariat, Tokyo Secretariat

Target inspection rate	25% annual inspection rate per country within 3 years of effective date (1982)	15% annual inspection rate per country within 3 years of effective date (1992)	25% annual regional inspection rate by the year 2000	15% annual inspection rate per country within 3 years of effective date (1996)
Governing body	Port State Control Committee	Port State Control Committee	Port State Control Committee	Port State Control Committee
Secretariat	Provided by the Netherlands Ministry of Transport and Public Works (Rijswijk)	Provided by Prefectura Naval Argentina (Buenos Aires)	Tokyo MOU Secretariat (Tokyo)	Hosted by Barbados
Database center	Centre administratif des affaires maritime (CAAM) (Saint-Malo)	Centro de informacion del acuerdo latinamerico (CIALA) (Buenos Aires)	Asia-Pacific Computerized Information System (APCIS) (Ottawa)	Caribbean Maritime Information Center
Official language	English, French	Spanish, Portuguese	English	English
Signed	January 26, 1982	November 5, 1992	December 2, 1993	August 8, 1996
Effective date	July 1, 1982	November 5, 1992	April 4, 1994	August 8, 1996, but signatories have 3 years to put an administration in place

BOX 2-1
U.S. Port-State Control Initiative

The USCG has established as a major part of its Port-State Control Initiative the goal of identifying and eliminating substandard foreign-flagged ships from U.S. waters while encouraging operators trading to the United States to adopt management philosophies that ensure compliance with accepted standards. The program pursues this goal by systematically targeting high-risk vessels for inspection.

The targeting scheme places ships in one of four categories:

Priority I = boarded at sea prior to being allowed into port
Priority II = boarded after entry into port but prior to cargo transfer or passenger embarkation
Priority III = boarded after entry into port without restrictions on cargo transfer or passenger embarkation
Priority IV = not boarded

Targeting criteria include type of vessel, accident record, owner history, flag history, classification society history or status, and boarding history, with special attention to previous interventions, incidents, and violations.

committee was unable to prove or disprove the impression shared by a number of members of the maritime community that the quality of the fleet calling on U.S. ports has improved. In most of the other areas discussed below, the committee encountered similar data problems when attempting to draw comparisons between pre- and post-OPA 90 periods.

The committee identified several unilateral efforts by foreign countries to improve ship safety and decrease the potential for oil pollution. Japan has initiated a system designed to discourage the use of older ships for trade to that country. Vessels more than 15 years of age are approved for trade only after careful vetting. Korea has indicated that it will discourage ships over the age of 20 except for those with high condition assessment program (CAP) ratings. Finland gives a rebate on port charges to ships with double-bottom, double-side, and full double-hull designs.

Enhanced Surveys and Classification Society Activities

Enhanced surveys of vessels that comply with guidelines developed by IMO have been mandated by MARPOL 13G and are included in USCG regulations. Such enhanced surveys are applicable to 5-year periodic, 2.5-year intermediate, and annual surveys. They entail increasingly strict inspections as vessels age and are intended to deter the operation of substandard ships that could result in oil spills due to structural failure.

The larger classification societies, notably members of the International Association of Classification Societies (IACS),[8] have begun aggressive programs to ensure that vessels under their classification meet or exceed present requirements. IACS members, who together classify 42,000 ships that account for 90 percent of the world's merchant ship tonnage, have established initiatives that reflect an increasing attention to safety, as well as a willingness to share information within the maritime community. These initiatives are intended to reduce the number of substandard ships and include (1) the sharing of information with third-party organizations having a legitimate interest in the maintenance of safe shipping standards and their application; (2) revised procedures for the suspension of class if special surveys, annual surveys, or recommendations or conditions of class fall overdue; (3) employment, control, certification, and training of surveyors; (4) a transfer of classification agreement whereby no ship can be accepted for membership in a member society unless it has addressed unresolved issues from its previous society; (5) coordination of classification surveys with port-state control surveys; and (6) enhanced surveys.

Industry Initiatives

Charterer Vetting Programs

Major charterers have developed sophisticated vetting (audit and inspection) programs that they use prior to chartering a vessel. Such programs include vessel inspections, consideration of flagging history, and ownership and management qualification requirements. Some vetting programs have been in place for a number of years, whereas others are relatively recent. Information from such programs is now available to other charterers, terminal operators, port authorities, and government agencies through the SIRE Ship Inspection Programme sponsored by the Oil Companies International Marine Forum (OCIMF) (see Box 2-2).

An indirect result of the adoption of OPA 90 has been an increased emphasis on safety in existing quality programs and an increased reliance among charterers on records of vessel adherence to these programs. Variations in the programs and their partly subjective nature precluded any assessment by the committee of their effect on the quality of the maritime oil transportation fleet.

---

[8]IACS members include the American Bureau of Shipping (ABS), Bureau Veritas, China Classification Society, Det Norske Veritas (DNV), Germanischer Lloyd, Korean Register of Shipping, Lloyd's Register of Shipping, Nippon Kaiji Kyokai, Polski Rejestr Statkow, Registro Italiano Navale, and Maritime Register of Shipping (Russia). The Croatian Register of Shipping and the Indian Register of Shipping are associate members.

BOX 2-2
Ship Inspection Report (SIRE) Program

OCIMF has recently begun a voluntary information sharing program designed to help improve ship safety and promote efficiency. This program takes advantage of the information that has been gathered by many of its members as part of their separate tanker vetting programs. Many member companies have been involved in tanker vetting for years, whereas others have become involved more recently.

SIRE is designed to provide a readily accessible pool of technical information concerning the condition and operational procedures of tankers. The information is available not only to OCIMF members but also to companies and organizations chartering oil tankers, bulk oil terminal operators, port and canal authorities, and government agencies having flag- or port-state responsibility for tanker safety.

The information contained in SIRE is provided by participating OCIMF members from the information that they gather for their own vessel vetting. One copy is sent to the operator of the vessel inspected and another copy is submitted to OCIMF for posting in a computerized database. The tanker operator has 14 days to submit written comments for inclusion in the database. The information then becomes available to third parties. OCIMF also maintains a computerized index that gives information about the availability of reports and tanker operator comments.

Most of the major oil companies that are members of OCIMF are voluntary program participants. Stated goals are (1) to expand the availability of tanker inspection information and, by so doing, enhance tanker safety with a consequent reduction in pollution; and (2) to reduce duplication of efforts by inspecting organizations and thus reduce the burden placed on tanker crews.

There is no set format for inspection reports, but many members base their reports on the OCIMF publication *Inspection Guidelines for Bulk Oil Carriers*. These guidelines address questions of vessel safety and pollution prevention, manning levels, certification and competency, mooring equipment, navigation procedures, general condition, and other safety-related items. Reports made available by OCIMF reflect all the information submitted by the inspecting party. The reports do not contain an overall tanker rating or any indication of the inspecting company's view of the ship's acceptability. It is left to the user to determine whether a vessel is acceptable for its intended use.

Quality Criteria for Older Ships

Major charterers, such as Vela of Saudi Arabia, Norwegian Statoil, Norsk Hydro, and the government of Kuwait, require ships exceeding 20 years of age to achieve high ratings in the CAP of the ABS, Det Norske Veritas, and Lloyd's classification societies. On the basis of stringent inspections of each vessel, CAP assigns ratings on a scale jointly agreed to by the classification societies.[9] Of significant interest to the United States is the fact that the fleet of ultralarge crude carriers (ULCCs) of more than 400,000 DWT, consisting of 25 ships used almost entirely in trade to the United States, has 20 ships that have gone through CAP with a Grade 1 rating.

VESSEL QUALITY

Despite the significant number of new or enhanced programs now in place that are intended to reduce the number of substandard vessels in use (thereby reducing oil pollution and increasing safety), the committee was not able to document conclusively any changes in the quality of the fleet calling on the United States since the promulgation of OPA 90. The subjective nature of much of the available data, the lack of reliable evaluation information and quality rating data, and the difficulty of establishing comparisons between datasets for different years all contributed to the committee's inability to make a definitive statement about vessel quality.

Nonetheless, many of the presentations made to the committee suggested that the quality of the fleet trading to the United States has improved in recent years, a view echoed by committee members who work in the maritime community. Possible reasons for the perceived improvement cited by presenters were (1) an increased awareness of potential liability;[10] (2) improvements in company operations; (3) increased familiarity with the harmful consequences of an oil spill as a result of mandatory planning and training in responding to spills; and (4) a greater emphasis placed on protecting the environment by several elements of the maritime community, including the IACS, protection and indemnity (P&I) clubs, port states, owners and operators, vessel crews, and individual U.S. states, including California, Texas, and Washington.

Although the lack of reliable and consistent data precluded any definitive assessment of changes in vessel quality and related safety, the committee anticipates that future assessments may be more conclusive. Data on the physical

[9]Grade 1 = very good; superior condition. Grade 2 = good; condition above average. Grade 3 = satisfactory; condition not necessitating repairs; equivalent to enhanced special survey. Grade 4 = unsatisfactory; repairs needed.

[10]Insurance for pollution liability up to $500 million is provided by protection and indemnity (P&I) clubs. Most tanker owners take advantage of commercial options that offer further coverage of up to $200 million.

condition of vessels, as well as on ships and owners with high incident records, are being collected in several of the initiatives described, notably the U.S. Port State Control Initiative and the charterer vetting programs. Such data might provide the basis for future analyses of vessel quality and safety.

FINDINGS

Finding 1. During the period of 1990 to 1995, compared to earlier five-year periods, there was a decline in the quantity of oil spilled from vessels in U.S. waters, together with a reduction in the number of spills of more than 100 gallons.

- Since 1990, there have been no single oil spills of greater than 1 million gallons from tank vessels in U.S. waters. Such incidents have dominated oil spill statistics over the past two decades; even one incident has the potential to change the overall statistical picture dramatically.
- The majority of all oil spills in U.S. waters between 1990 and 1995 involved single-hull barges. Barges accounted for approximately 50 percent of the total oil spilled during this period, with tankers accounting for approximately 10 percent. Fishing vessels, passenger vessels, dry cargo carriers, and other types of ships accounted for the remaining 38 percent.

Finding 2. Because of the timing of its implementation, Section 4115 of OPA 90 did not have a major influence on the observed decline in the volume of oil spilled from vessels in U.S. waters between 1990 and 1995. The double-hull provisions are just beginning to be implemented, and during the period of study, the operational measures aimed at reducing the outflow of oil following incidents involving single-hull tank vessels had not been fully implemented.

Finding 3. Major international programs such as MARPOL 13F and 13G, port-state initiatives, and enhanced inspections by classification societies, are just now coming into effect and are not, in and of themselves, the reason for the reduction in oil spills for the period 1991 to 1995. However, some anecdotal information suggests that there has been an improvement in the quality of the fleet calling on U.S. ports because of an increased awareness of liability obligations and the cost of pollution liability insurance, enhanced planning and training for oil spills, improved audit and inspection programs by charterers and terminal operators, policies and procedures adopted by fleet or vessel operators, and increased rules imposed by port states (such as the U.S. Port State Control Initiative) and by individual U.S. states.

Finding 4. The passage of OPA 90 was a catalyst for international action, resulting in the addition of Regulations 13F and 13G to MARPOL and the worldwide adoption of a double-hull mandate for tank vessels.

Finding 5. The committee found the USCG oil spill database to be virtually inaccessible to the general public and of uneven quality from district to district. Major difficulties exist in the management of this important source of information on trends in and causes of oil spills, making specific analyses difficult. Extraordinary efforts were needed to obtain adequate data for the analysis of oil spill trends reported in Finding 1.

Finding 6. The USCG databases on port-state inspections for 1990 and 1995 cannot be used to make comparisons of fleet quality because the data for 1990 do not include information on the total number and type of vessels inspected. A new approach to port-state inspection now being used by the USCG could, over time, provide suitable data for future comparisons if the database is appropriately maintained by the USCG and includes information such as the total number of vessels making port calls and the number and type of vessels inspected.

Finding 7. Probabilistic outflow analysis of existing vessel designs (see Chapter 6) indicates that the complete conversion of the maritime oil transportation fleet of tankers and barges to effectively designed double hulls is expected to result in significantly improved protection of the marine environment. Reductions are anticipated in both the number of spills and the volume of oil spilled.

REFERENCES

Federal Register. 1994. Emergency Lightering Equipment and Advanced Notice of Arrival Requirements for Existing Tank Vessels Without Double Hulls. Final Rule. FR 59(150):40186–40189, August 5.

Federal Register. 1995. Structural Measures to Reduce Oil Spills from Existing Tank Vessels Without Double Hulls. Supplemental Notice of Proposed Rule Making. FR 60(249):67226–67251, December 26.

Federal Register. 1996. Operational Measures to Reduce Oil Spills from Existing Tank Vessels Without Double Hulls. Final Rule. FR 61(147):39770–39794, July 30.

Federal Register. 1997a. Structural Measures to Reduce Oil Spills from Existing Tank Vessels Without Double Hulls. Final Rule. FR 62(7):1622–1637, January 10.

Federal Register. 1997b. Operational Measures to Reduce Oil Spills from Existing Tank Vessels Without Double Hulls. Final Rule; Response to Petitions for Rulemaking. FR 62 (184):49603–49608, September 23.

Herbert Engineering Corporation. 1996. Comparative Study of Double-Hull vs. Single-Hull Tankers. Background paper prepared for the Marine Board Committee on OPA 90 (Section 4115) Implementation Review. San Francisco, Calif.: Herbert Engineering Corporation.

Parliament of the Commonwealth of Australia. 1992. Ships of Shame: Inquiry into Ship Safety. Report from the House of Representatives Standing Committee on Transport, Communications and Infrastructure. Canberra: Australian Government Publishing Service.

U.S. Coast Guard (USCG). 1993. OPA 90 Deepwater Ports Study. Washington, D.C.: U.S. Coast Guard Office of Marine Safety, Security and Environmental Protection.

3

Operational Makeup of the Maritime Oil Transportation Industry

This chapter identifies the nature and extent of changes in tank vessel ownership and use in the maritime oil transportation industry since 1990. The committee analyzed the following factors:

- changes in the size distribution of vessels used and related trading patterns, including the use of the deepwater port and lightering zones
- changes in the age distribution of the world and U.S. trading fleets and scrapping patterns
- changes in vessel ownership, including trends in sales and transfers

To assess the impact of the Oil Pollution Act of 1990 (P.L. 101-380) (OPA 90) since its enactment, the committee compared data for 1990 (the last pre-OPA 90 year) with data for 1994 or 1995. Where possible, projections for the period 1995 through 2015 were analyzed to provide an indication of likely changes in the operational makeup of the fleet during the phaseout period for single-hull tank vessels.

The analysis treats the international fleet and the domestic (Jones Act) fleet separately because of their different trading patterns and composition. The international fleet carries crude oil and finished products to the United States from foreign sources, whereas the Jones Act fleet trades almost exclusively between U.S. ports, with an occasional cargo for foreign aid or for the Military Sealift Command going abroad. In addition, the composition of the two fleets is substantially different. The average age of vessels in the Jones Act fleet is greater than that of vessels in international trade.

VESSEL SIZE AND TRADING PATTERNS

The makeup of the international fleet trading to the United States is determined primarily by U.S. requirements for oil imports and the geographical distribution of supplier nations. Most U.S. oil imports are crude oil. In 1994, for example, U.S. oil imports totaled 8.1 million barrels per day (MBD), of which 7.0 MBD were crude oil. The predominance of crude oil imports is expected to continue through 2015 (EIA, 1996).

The longer the distance that crude oil must travel before reaching the United States, the larger the tanker used is likely to be. The selection of tanker size is determined by logistics and economics. The economics of transporting long-haul crude usually favor the use of very large crude carriers (VLCCs). Many crude oil loading terminals overseas are located at ports that can accommodate fully loaded VLCCs. However, there are no ports on the U.S. east or Gulf coasts deep enough to accommodate such ships when fully loaded. Hence, there is a need for deepwater ports and lightering zones. The economic decision regarding tanker size is often based on a comparison between the higher chartering cost per barrel of crude oil shipped in small tankers and the lower chartering cost per barrel of oil shipped in larger tankers, combined with lightering costs or the cost of unloading at the Louisiana Offshore Oil Port (LOOP).[1] Short-haul crude oil imports (e.g., from the Caribbean) are generally shipped in smaller tankers that either sail directly into U.S. ports or can be partially lightered so that most of the cargo remains on the tanker, which is then moved into port and unloaded.[2]

Trends between 1990 and 1994

The amount of crude oil and finished products imported into the United States by water increased by 19 percent from 1990 to 1994 (see Table 3-1).[3] The Atlantic coast showed a decline of 6 percent, the Gulf coast an increase of 28 percent, and the Pacific coast an increase of 69 percent. The committee found that these changes in import patterns caused a change in the size distribution of oil tankers trading to the United States between 1990 and 1994. In 1990, the large crude carriers that use the LOOP or lightering areas in the Gulf of Mexico were the largest group in the Gulf compared to all others that entered Gulf ports (with or without local lightering), including crude carriers, fuel oil carriers, and product carriers. By 1994, as shown in Table 3-1, more cargo was being transported to the Gulf in 80,000 to 150,000-deadweight ton (DWT) vessels.

---

[1]Lightering and off-loading at deepwater ports are generally limited to crude oil. Unfinished oil material and finished petroleum products are transported in smaller ships to facilitate unloading at product terminals.

[2]Most crude oil imports from Canada into the United States flow via pipeline.

[3]Unless otherwise noted, tables and figures in this chapter were developed from data supplied by the Institute of Shipping Analysis of Göteborg, Sweden, based on data from Lloyd's Maritime Information Services.

TABLE 3-1 Change in Tonnage, by Coast and Vessel Size, 1990–1994

	Atlantic			Gulf			Pacific			Total		
Size Category[a]	1990 (10^6 tons)	1994 (10^6 tons)	Change (%)	1990 (10^6 tons)	1994 (10^6 tons)	Change (%)	1990 (10^6 tons)	1994 (10^6 tons)	Change (%)	1990 (10^6 tons)	1994 (10^6 tons)	Change (%)
10–40	14.5	5.9	–59	2.7	1.8	–33	1.5	1.2	–20	18.8	8.9	–53
40–80	28.3	20.7	–27	23.9	26.4	10	2.9	6.6	128	55.0	53.7	–2
80–150	61.7	72.8	18	77.8	122.2	57	6.8	9.6	41	146.4	204.6	40
150+	4.9	3.6	–27	102.6	114.8	11	2.4	5.6	133	109.9	123.9	13
Total	109.5	102.9	–6	206.9	265.2	28	13.6	23.0	69	330.1	391.2	19

[a]For the purposes of the committee's analysis, oil tankers were categorized in four size ranges: 10,000 to 39,999 DWT (10–40); 40,000 to 79,999 DWT (40–80); 80,000 to 149,999 DWT (80–150); and 150,000 DWT or more (150+).

Table 3-1 also shows that the number of product tankers (vessels between 10,000 and 40,000 DWT) unloading in U.S. ports decreased by 53 percent, with decreases on all coasts but notably on the Atlantic. The sharp declines in tonnage carried in these vessels to the Gulf and Pacific coasts do not so much reflect a reduction in the movement of products and small lots of crude, but rather the increasing use of vessels of 40,000 to 45,000 DWT to carry products. On the Atlantic coast, there was a decrease in imports of products and a switch from small- and medium-sized vessels to tankers of more than 80,000 DWT to carry crude oil, but there was little change in volume.

Through the years, 40,000 to 80,000 DWT tankers have played a significant role in transporting oil to the Atlantic and Gulf coasts, but the amount of cargo carried to the United States in this size vessel decreased by 2 percent between 1990 and 1994. Only the Atlantic coast showed an actual decrease in cargo (minus 27 percent), with the Gulf showing an increase of 10 percent and the Pacific coast more than doubling.

The greatest increase in tonnage carried is seen in the 80,000 to 150,000 DWT category. This is explained partly by increased crude oil imports from South America and the Caribbean. This category of vessels is well suited to the short- and medium-haul trade, because they are able to deliver their cargo directly to the dock, unlike the larger vessels from long-haul destinations such as the Arabian Gulf.

The largest size category, 150,000+ DWT, also increased in 1994 versus 1990, although the increase was less than the overall increase in total imports. This reflected a change in the sources of crude oil rather than the influence of OPA 90 regulations.

Projected Trends between 1995 and 2015

There are indications that the United States will need more petroleum imports in 2015 than in 1994; projections suggest that a major share of the increase will be crude (EIA, 1996). Generally, oil imports from relatively close locations in the Western Hemisphere[4] have a logistical cost advantage over imports from more distant locations. Thus, crude oil from the Western Hemisphere is likely to be the first choice of U.S. importers, although its availability may be restricted. In 1994, U.S. crude oil imports from the Western Hemisphere were 3.7 MBD. However, this figure is expected to decrease to 3.4 MBD by 2015 as demand in the Western Hemisphere outside the United States increases faster than production (EIA, 1996).

The limited availability of short-haul crude is expected to result in an increase in imports from more distant locations, such as the Arabian Gulf. Analyses

[4]In 1994, about 75 percent of the 10.9 MBD of oil produced in the Western Hemisphere outside the United States was produced in Canada, Mexico, and Venezuela (EIA, 1996).

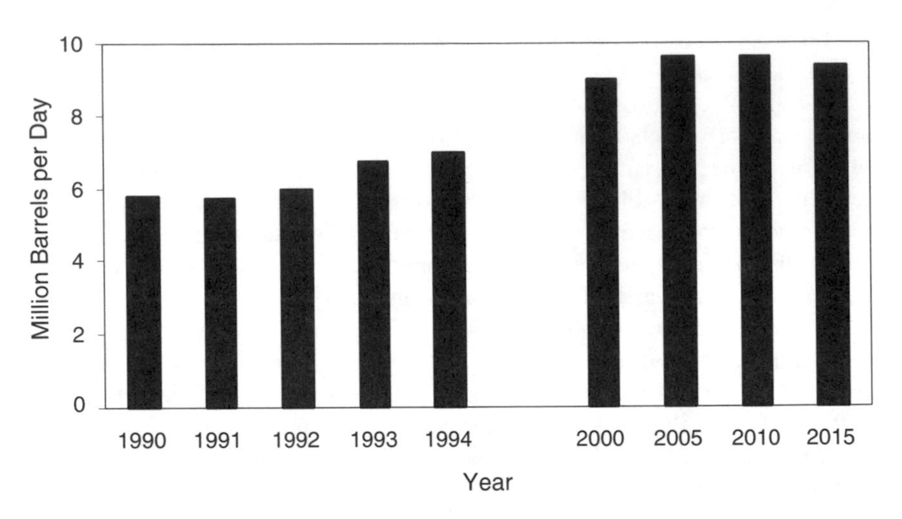

FIGURE 3-1 Projections of U.S. crude oil imports through 2015. Sources: 1990–1994 data from EIA, 1994; 2000–2015 data from EIA, 1996.

of U.S. Department of Energy projections indicate that long-haul crude imports into the United States are likely to increase by more that 2.5 MBD between 1994 and 2015 (Figure 3-1). Studies by the Petroleum Industry Research Foundation, Inc. (PIRA, 1992) and Wilson, Gillette & Co. (Wilson, Gillette, 1994) support the same conclusion.

The projected increase in long-haul crude imports will mean an increase in the number of large tankers trading to the United States. Long-haul crude oil imports arriving at the Gulf Coast are projected to increase from 2.4 MBD in 1994 to 5.1 MBD in 2015 (see Figure 3-2). The relative costs and availability of large single-hull and double-hull tankers will determine, in part, which type of vessel is used. In addition, the provisions of OPA 90 (Section 4115) are likely to have an influence on the composition of the VLCC fleet trading to the United States.

Currently there is only one deepwater port on the U.S. Gulf Coast. Figure 3-2 assumes a continuation of crude oil import operations at this facility following historical trends. Perceived economics will ultimately determine the potential for expanding this site or building new deepwater ports. Currently, two additional deepwater ports are under consideration for the U.S. Gulf Coast. Designated lightering areas can handle more oil than LOOP (see Figure 3-2), and some consideration is being given to designating more such areas.

Effects of OPA 90

The committee was unable to isolate any changes in trading patterns and sizes of vessels calling at U.S. ports between 1990 and 1994 that could be clearly

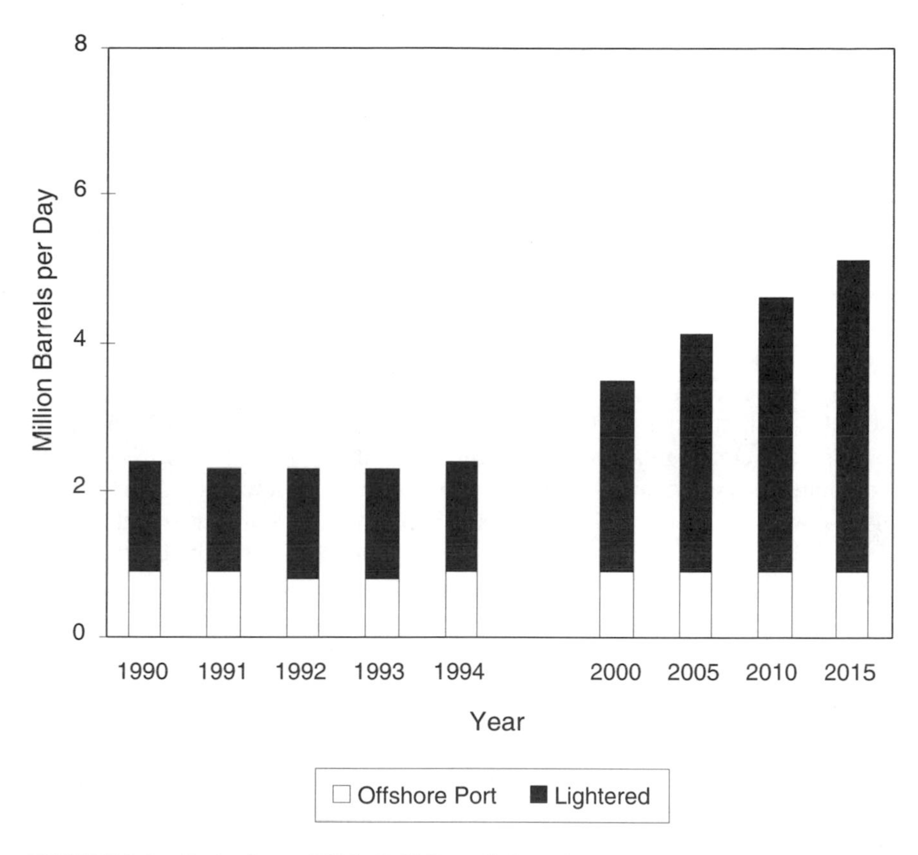

FIGURE 3-2 Projections of U.S. Gulf Coast lightered and deepwater port crude oil imports through 2015. Sources: 1990–1994 data from EIA, 1994; 2000–2015 data from EIA, 1996.

attributed to OPA 90 or to the International Convention for the Prevention of Pollution from Ships, adopted in 1973 and amended in 1978, Regulations 13F and 13G of Annex I (13F and 13G). However, some significant changes in trading patterns are anticipated over the next 20 years as single-hull vessels are phased out in accordance with OPA 90 and International Maritime Organization (IMO) requirements.

Smaller single-hull tankers that move directly into port or are partially lightered will be phased out by 2010 in accordance with OPA 90.[5] In contrast, the deepwater port and lightering zone exemption to OPA 90 will allow large single-hull tankers to unload through 2015 at LOOP or in lightering zones designated by the

[5]Some tankers with a double bottom or double sides may be allowed to operate through 2015, depending on their age.

Coast Guard. The availability of such areas will extend the period for using large single-hull tankers in U.S. waters by five years compared to the time when other single-hull tankers must be phased out.

The significantly lower capital costs of older single-hull tankers will allow them to operate profitably even with lower charter rates than the more desirable double-hull tankers, and charterers are likely to take advantage of the lower rates. Even after taking into account the reduction in cargo and revenue due to adopting hydrostatically balanced loading (HBL), estimated at 8 percent, and the special survey costs necessary to extend their lives from 25 to 30 years (see Chapter 4), pre-MARPOL tankers should be able to operate at a profit at rates significantly lower than new double-hull tankers. Thus, the combination of the deepwater port and lightering zone exemption and potentially lower operating costs suggests that some single-hull tankers will continue to operate in U.S. waters until 2015. In this one respect, U.S. regulations are less restrictive than international rules, and the committee assumes that given favorable economics, shipowners will take maximum advantage of the exemption for single-hull VLCCs built between 1985 and 1993. Thus, some single-hull VLCCs may continue to trade to the United States until age 30.

The continued operation of older single-hull vessels to the U.S. deepwater port and lightering zones has caused some questions to be raised about the quality of vessels trading to the United States. In addition, the efforts of other countries to attract modern vessels with desirable safety features have heightened concern in some quarters that the quality of the international fleet trading to the United States may be compromised unless appropriate measures are taken. The possible introduction of incentives for early retirement of single-hull vessels trading to the United States has been raised in the course of congressional hearings[6] and in the State of California.

AGE DISTRIBUTION AND SCRAPPING PATTERNS

The committee's assessment of the operational makeup of the maritime oil transportation industry focuses on the part of the international oil transportation fleet that trades to the United States. However, it should be recognized that subject to regulatory and other restrictions, the entire world fleet is potentially able to trade to the United States. In practice it is not always possible to identify with confidence which vessels will come to the United States and which will not. Thus, even if the same numbers of vessels were to trade to the United States for two consecutive years, the actual vessels involved would not all be the same. As noted in Chapter 2, there are programs in various countries to encourage the use of newer vessels in their waters. These programs will increase competition for

[6]Hearings of the Committee on Environment and Public Works, U.S. Senate, on S. 1730, a bill to amend the Oil Spill Prevention and Response Improvement Act, Washington, D.C., June 4, 1996.

newer vessels in the world market and may influence the age distribution of that portion of the world fleet trading to the United States. Therefore, in the context of its analysis of age distribution and scrapping patterns, the committee considered it appropriate to include a discussion of the features of the total world fleet before addressing the portion that trades to the United States.

World Fleet

Table 3-2 shows the change in the composition of the world fleet from 1990 to 1994 in terms of construction type—namely, double-hull tankers, single-hull tankers built prior to 1980 and mostly of pre-MARPOL configuration, and single-hull tankers built after 1980 (MARPOL tankers). The proportion of double-hull tankers increased from 4.0 percent in 1990 to 10.1 percent in 1994.

The scrapping profile for tankers in the world fleet over the period 1990 to 1995 is shown in Figure 3-3. A distinction is drawn between ships less than and more than 150,000 DWT. Data on the numbers of vessels scrapped during the same period are provided in Table 3-3. The average age at scrapping of larger tankers during the period was approximately 20 years, whereas for smaller tankers it was about 25 years (Figure 3-4). Data for the smaller tankers indicate a slight decrease in average age at scrapping through 1994, followed by a slight increase in 1995. Data for the 150,000 DWT and greater category show little variation in average scrapping age during the years of significant scrapping (i.e., 1992–1995).

Figure 3-5 shows the relationship between freight rates and the total tonnage of vessels scrapped from the world fleet between 1982 and 1995. Only a small number of ships were scrapped in 1990 and 1991, which were years of relatively high freight rates, whereas a much larger number of vessels were scrapped in 1992, a year of very low rates. As illustrated in Figure 3-5, scrapping rates appear to be influenced by economics, although there is some lag between cause and effect.

TABLE 3-2 Change in Composition of World Fleet between 1990 and 1994, by Hull Type as Percentage of Total Tonnage[a]

Year	Double Hull	Single Hull Built before 1980	Single Hull Built 1980 or Later	Total Fleet
1990	4.0	65.1	30.9	100
1994	10.1	51.2	38.7	100

[a]In 1990 the world fleet comprised 3,305 vessels of greater than 10,000 DWT; in 1994 there were 3,380 such vessels.

Sources: Clarkson Research, 1990, 1995; Tanker Advisory Center, 1991, 1995.

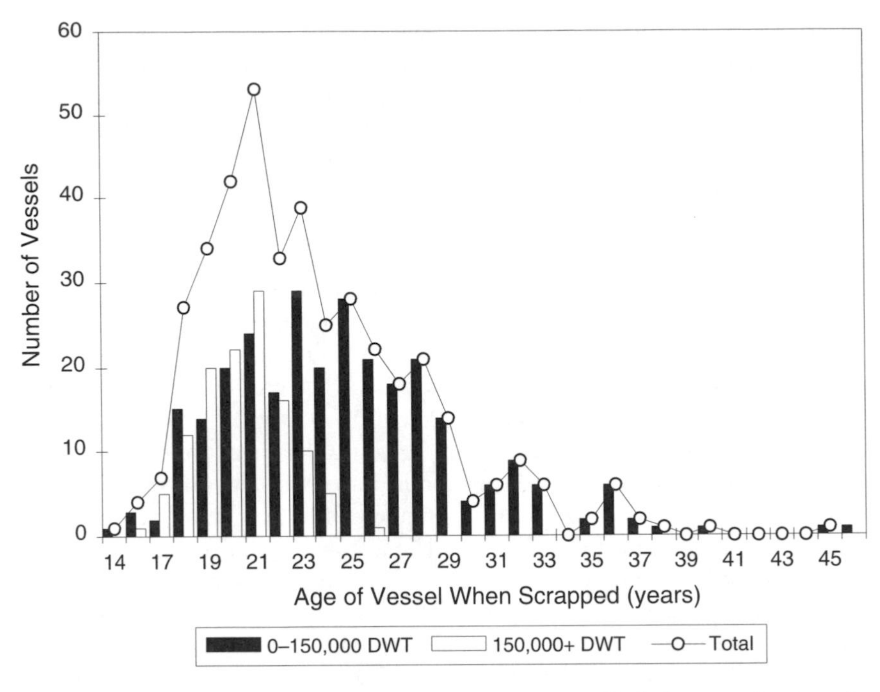

FIGURE 3-3 Scrapping profile for the world fleet, 1990–1995. Sources: Clarkson Research, 1991–1995, 1996a,b; Drewry, 1994; Tanker Advisory Center, 1996.

The most pronounced effect of Section 4115 of OPA 90 and MARPOL 13F and 13G on the world fleet so far has been an increase in the number of double-hull vessels. This trend is expected to continue during the mandatory phaseout periods. There is little indication that scrapping has been influenced by Section 4115. This observation is not unexpected; the required phaseout age is greater

TABLE 3-3 Tankers Scrapped per Year from World Fleet, 1990–1995

	Deadweight Tonnage		
Year	10,000 to 150,000	150,000 or More	Total
1990	11	1	12
1991	29	1	30
1992	70	20	90
1993	91	31	122
1994	52	37	89
1995	64	33	97
Total	317	123	440

Sources: Clarkson Research, 1991–1995, 1996a,b; Drewry, 1994, 1996.

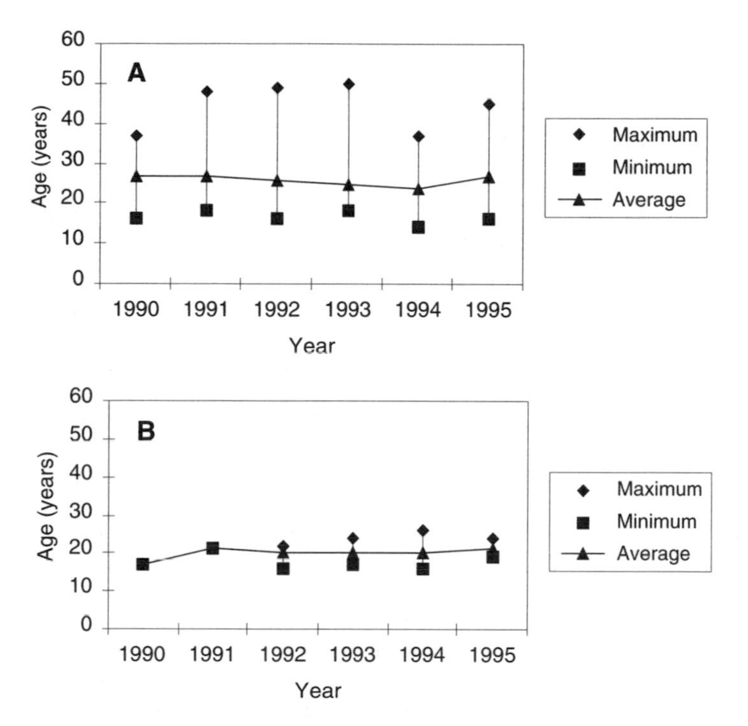

FIGURE 3-4 Age of tankers scrapped from world fleet, 1990–1995. (a) < 150,000 DWT; (b) ≥ 150,000 DWT. Sources: Clarkson Research, 1990–1995, 1996a,b.

than the normal age at scrapping. No change in scrapping patterns is expected until the early part of the next century, particularly for smaller tankers.

Although Section 4115 and MARPOL 13F and 13G are expected to increase the scrapping of tankers in the world fleet in the future, they may not produce the massive scrapping anticipated by some observers. Section 4115 will have little impact on the retirement of large single-hull tankers (150,000 DWT or more) that are suitable for unloading within the lightering zones or at the deepwater port (see Figure 3-6). Moreover, tankers built with segregated ballast tanks (SBTs) during the boom of the 1970s can continue to operate until age 30 under IMO rules. Those without SBTs can trade through age 30 if they are fitted with SBTs or use HBL. The latter option involves a loss of capacity of between 5 and 13 percent for larger vessels.

In summary, the committee's analysis indicates that the majority of small- and medium-sized tankers in the world fleet will be scrapped before they reach the maximum age permitted by the regulations, unless historical scrapping patterns change significantly. Older vessels will be competing in a market where replacement vessels will be significantly more expensive to build, albeit more efficient to operate. Well-maintained and efficient older vessels should be able to

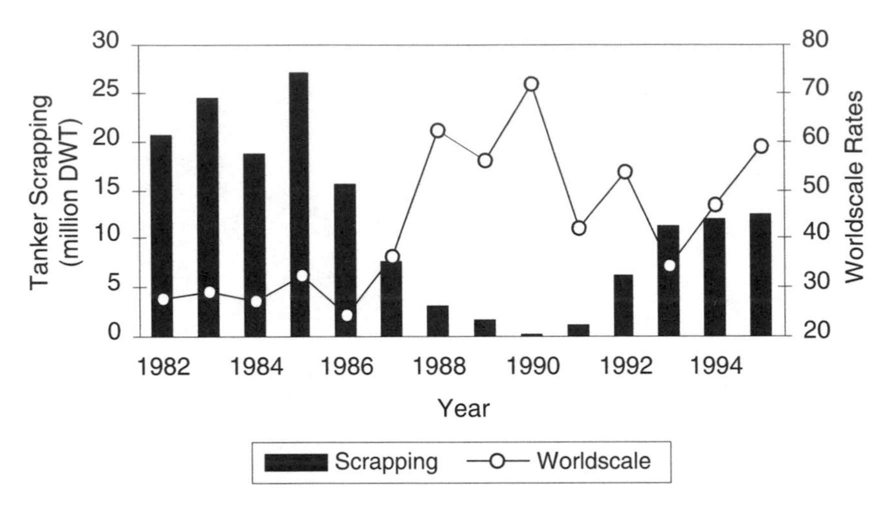

FIGURE 3-5 Freight rates and total tonnage scrapped from world fleet, 1982–1995. Note: Worldscale is the common designation for the New Worldwide Tanker Nominal Freight Rate Scale applying to the carriage of oil in bulk. The Worldscale Schedule is published annually by the Worldscale Association (London) and Worldscale Association (New York). Worldscale (WS) is a series of calculated costs for the voyages, as listed in the schedule, between designated ports for the Worldscale "standard" tanker (75,000 DWT). The Worldscale calculated voyage cost, in dollars per metric ton of cargo carried, is a benchmark used in negotiations between vessel owners or operators and charterers. Voyage charter rates are typically agreed to in terms of percentage of Worldscale (i.e., a charter at 125WS is for 125 percent of the calculated voyage cost of the standard tanker for that voyage). Sources: Clarkson Research, 1990–1995, 1996a; Jacobs & Partners, 1995.

earn a handsome return, thereby providing an incentive not to scrap. However, many vessels built during the boom of the 1970s are now reaching ages in the historical scrapping range, and the large number of these vessels may lead to a different scrapping scenario from that seen previously. The owners and operators of larger tankers are likely to take advantage of the lightering zone and deepwater port exemption to OPA 90, together with the HBL option allowed under IMO rules, and operate their vessels until age 30.

International Fleet Trading to the United States

Age and Scrapping Profiles

The average age of the U.S. trading fleet fell by 1.12 years from 1990 to 1994, even though the average age of the world fleet rose during the same period (see Table 3-4). Inspection of the age profile in Figure 3-7 indicates that the concentration of tankers trading to the United States in 1994 peaked in the

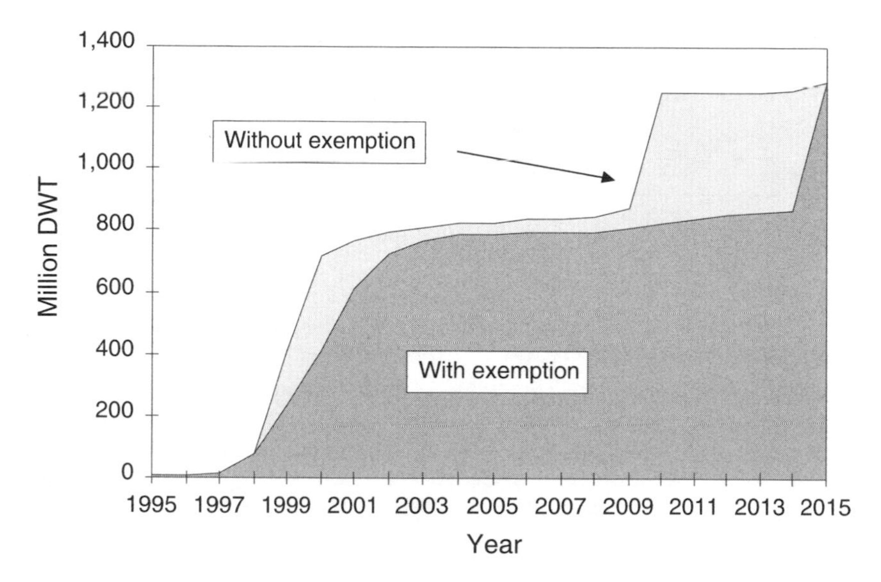

FIGURE 3-6 Deletions from world fleet due to OPA 90 and MARPOL, with and without lightering exemption for vessels of more than 150,000 DWT. Source: Navigistics, 1996.

youngest age group (0 to 4 years, built 1990 to 1994) and the fourth age group 15 to 19 years, built 1975 to 1979), with lower concentrations in intermediate age groups. Only 10 percent of total tonnage was carried in vessels more than 20 years of age, with 1 percent carried in vessels 25 years of age or older. The age profile was somewhat different in 1990, when the most common ages were 10 to 14 (built 1976 to 1980) and 15 to 19 years (built 1971 to 1975).

The explanation of these profiles lies in the economics of the market in the 1970s and 1980s. After 1976, the level of construction fell rapidly following the boom of the mid-1970s. Construction did not recover for some years, with relatively small expansions in the mid-1980s and after 1988. Thus, there is a "baby bust" generation, which is particularly noticeable in the 1981 to 1985 period. The latest wave of newbuildings, dating from 1988, had not yet made its mark on the youngest age bracket in the 1990 profile.

Examination of the estimated scrapping profile for vessels trading to the United States (Figure 3-8) indicates that for vessels built before 1978, scrapping typically occurred between 17 and 23 years of age. Only 6 percent of ships were still trading by age 23. The average life for tankers built in the 1970s has been about 20 years.[7] There is no doubt that the economic crises of the period contributed to this relatively short life. Maintenance was not kept up to optimum levels,

[7]There are some notable exceptions, including Jones Act ships whose cost of construction is so high that an extended life is economically justified and some others that were built under a philosophy of "heavy scantlings-intensive maintenance-long useful life."

TABLE 3-4 Changes in Age of U.S. Trading Fleet and World Fleet

	Average Age					
	U.S. Trading Fleet			World Fleet		
Age Range	1990	1994	Change	1990	1994	Change
0–4	2.20	2.03	–0.17	3.15	2.91	–0.24
5–9	7.54	6.79	–0.75	8.33	8.04	–0.29
10–14	12.62	12.36	–0.25	13.50	13.10	–0.40
15–19	16.01	17.57	1.56	17.31	18.55	+1.22
20–24	21.39	20.68	–0.70	22.60	22.08	–0.52
25 and older	35.00	34.05	–0.95	32.30	33.12	+0.82
Total	11.76	10.64	–1.12	14.08	14.73	+0.65

NOTE: Vessel age has been averaged both within the age band and overall for each ownership group.

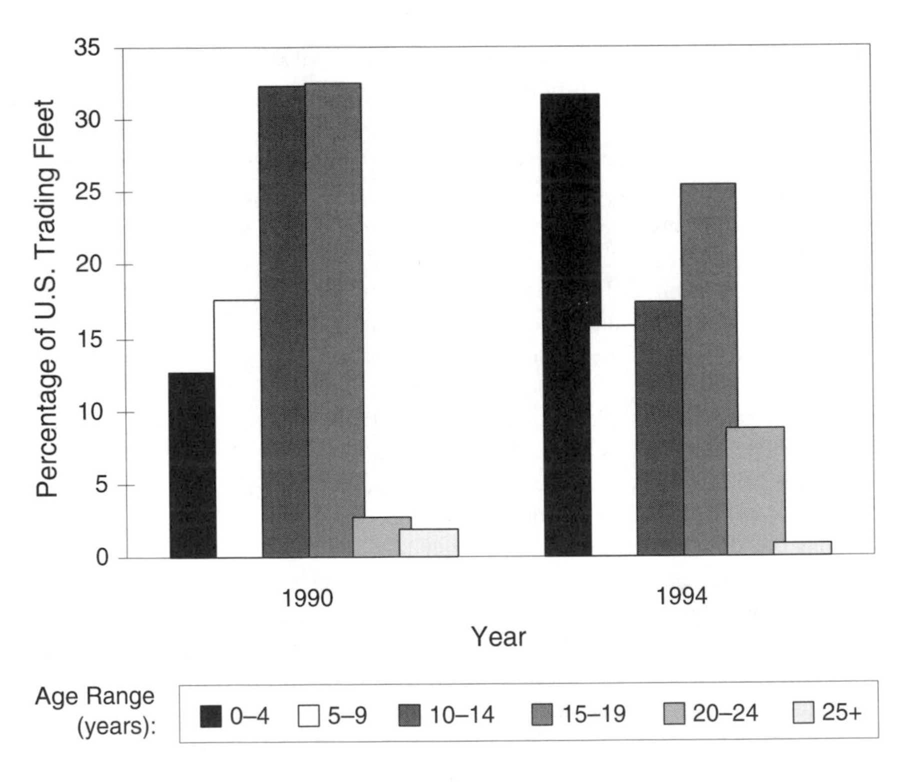

FIGURE 3-7 Tonnage carried by vessels trading to the United States by age of vessel for 1990 and 1994.

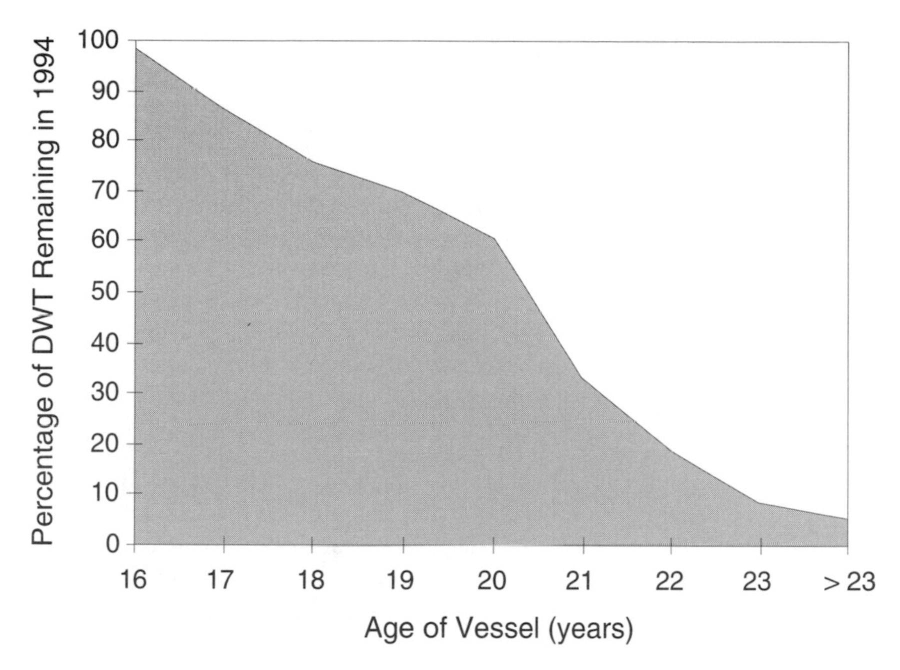

FIGURE 3-8 Estimated scrapping profile for tankers trading to the United States. Note: Figure shows percentage of the fleet built in years up to 1978 that remained in existence as of March 1995. The reduction in number of vessels remaining reflects mainly demolition, except for a few total losses, and is provided as a guide to useful life expectancy. Sources: Drewry, 1994; Tanker Advisory Center, 1995

and pre-MARPOL ships carried both cargo and ballast in the same unprotected tanks, which led to corrosion. Since about 1980 the use of SBTs has become common, and more recently the use of internal coatings (either partial on large ships or total on product carriers) has increased life expectancy, but these characteristics are not reflected in statistics on vessel age because the vessels have not yet reached the age of 20.

Port Calls and Cargo Discharge

The age characteristics of the international fleet trading to the United States can also be subdivided on a geographical basis (Table 3-5). The average age of the smaller tankers calling on the Atlantic coast increased from 1990 to 1994, whereas the average age of larger vessels decreased. Overall, the average age of vessels calling on the Atlantic coast decreased. The same pattern was found on the Gulf and Pacific coasts, with the average age of larger vessels decreasing and that of smaller vessels either showing little change or increasing. The overall

TABLE 3-5 Comparison of Average Age by Coast and Size Category

Size Category	Atlantic			Gulf			Pacific		
	1990	1994	Change	1990	1994	Change	1990	1994	Change
10–40	15.9	16.8	0.9	14.5	14.6	0.1	8.4	14.2	5.8
40–80	9.3	11.4	2.1	11.7	12.7	1.0	9.6	10.6	1.0
80–150	11.1	8.6	–2.5	11.1	7.8	–3.3	8.1	6.6	–1.5
150+	14.2	10.4	–3.8	14.4	13.9	–0.5	14.7	2.2	–12.5
Total	11.9	10.6	–1.3	11.9	10.1	–1.8	9.3	8.6	–0.7

average age of tankers calling on the Gulf coast decreased by 1.8 years; for the Pacific coast the reduction was 0.7 year.

Data on the number of port calls and tonnage carried per vessel in different age ranges are provided in Table 3-6. Comparison of the tonnage delivered by age of vessel in 1990 and 1994 indicates that vessels between 0 and 4 years and 20 to 24 years of age increased their amount of discharged oil significantly. The increased presence of newer vessels reflects the rise in new deliveries between 1990 and 1994 and the entry of these new-generation vessels in trade to the United States. The increase in the 20- to 24-year age bracket is indicative of the many vessels delivered during the mid-1970s construction boom.

The committee's analysis of the age distribution and scrapping patterns of the international fleet trading to the United States did not reveal any effects that can be attributed to OPA 90.

VESSEL OWNERSHIP, SALES, AND TRANSFERS

Ownership trends in the world fleet are addressed prior to an analysis of ownership changes in the international fleet trading to the United States. Different trends in the world and U.S. trading fleets since 1990 could indicate that OPA 90 had an impact on ownership. However, changes in the sources of and demand for crude oil worldwide and associated economic factors could also cause differences.

For the purposes of the committee's analysis, vessel ownership was subdivided into three categories: oil companies, government (including government-owned oil companies),[8] and independent owners. Unlike the first two groups, where the national identity of the owner is clear, it is not always possible to identify the nationality of an owner of an independently owned tanker. The flag

[8]In addition to government-owned fleets clearly identifiable as such, the committee also considered the following organizations to be government owned: Vela, Kuwait Oil, Petronard, Soponata, and Saudi Aramco.

TABLE 3-6 Average Size, Tonnage Carried, and Number of Port Calls, by Age of Vessel in the U.S. Trading Fleet for 1990 and 1994

Age Range (years)	1990					1994				
	Port Calls	Vessels	Tonnage Carried (10^6 tons)	Average Port Calls per Vessel	Average Tonnage per Vessel (10^3 DWT)	Port Calls	Vessels	Tonnage Carried (10^6 tons)	Average Port Calls per Vessel	Average Tonnage per Vessel (10^3 DWT)
0–4	469	160	41.7	2.9	88,900	1,031	336	123.7	3.1	120,000
5–9	816	255	58.4	3.2	71,600	715	211	61.6	3.4	86,200
10–14	910	341	106.9	2.7	117,500	908	267	68.5	3.4	75,400
15–19	768	338	107.7	2.3	140,200	645	280	100.0	2.3	155,000
20–24	115	31	9.2	3.7	80,000	205	82	34.2	2.5	166,800
25 and older	129	25	6.2	5.2	48,100	69	21	3.3	3.3	47,800
Total	3,207	1,150	330.1	2.8	102,900	3,573	1,197	391.2	2.99	109,500

and citizenship of the owner of record are known but do not necessarily identify who holds beneficial or controlling interests in a vessel.

World Fleet

During the period 1990 to 1994, independent ownership in the world fleet grew, mainly at the expense of oil company ownership. The number of government-owned ships also dropped but not to the extent found in the oil company sector.

There are clear trends in the pattern of vessel sales for the world fleet as a whole from 1990 to 1994, as summarized in Table 3-7. Although government fleets bought approximately the same number of tankers as they sold over the period, the pattern varied from 1990, when they were net buyers, to 1994, when they became net sellers. Independent fleet owners had a net surplus of purchases over sales of almost 10 percent, whereas oil companies were net sellers in all years from 1990 to 1994. It is important to note, however, that this pattern among the oil companies started before the enactment of OPA 90. The oil companies' share of the world fleet has been decreasing steadily over the past 20 years. Recent changes have included the sale of its fleet by Texaco in favor of an alliance with Stena and Exxon's sale of its VLCC fleet and switch to using charters.

Based on information presented to the committee by industry representatives (see Appendix D), it is apparent that several factors are responsible for the reduction in size of the oil company fleet. In particular, low tanker charter rates and high owner liability have caused many oil companies to deploy their capital in other ways. The committee was unable to establish the relative importance of these two factors.

International Fleet Trading to the United States

Between 1990 and 1994 the total tonnage of the international fleet trading to the United States increased by 18 percent, as shown in Figure 3-9. Over the same

TABLE 3-7 World Tanker Sales, 1990 to 1994

	Seller			
Buyer	Government	Independent	Oil Company	Total Sales
Government	24	59	5	88
Independent	69	852	82	1,003
Oil company	0	12	6	18
Total	93	923	93	1,109

Sources: Clarkson Research, 1990–1995, 1996a,b; Drewry, 1994, 1995; Tanker Advisory Center, 1996.

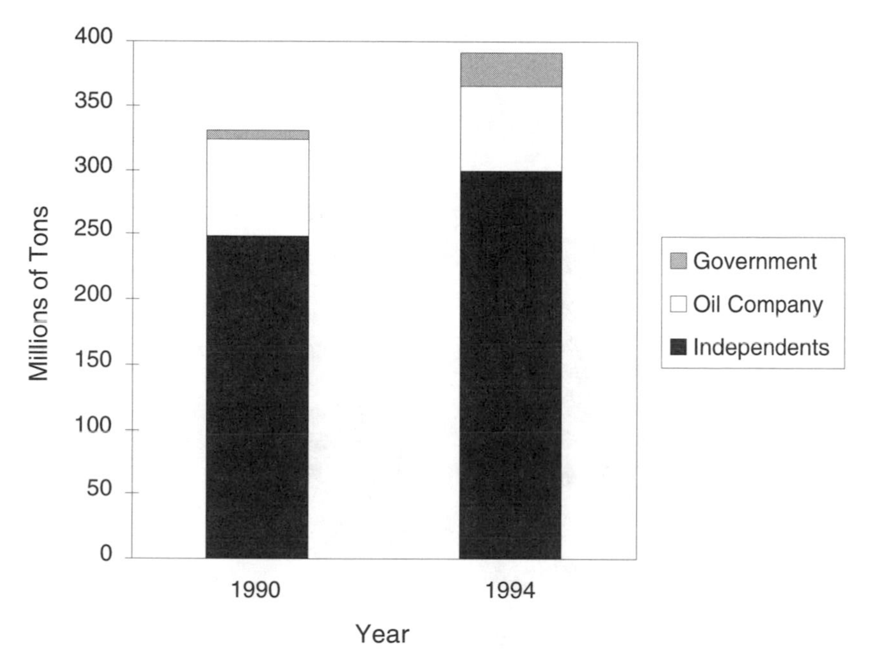

FIGURE 3-9 Changes in tonnage, by ownership category, for U.S. trading fleet between 1990 and 1994.

period the number of port calls increased only 11 percent (see Figure 3-10), indicating that the average size of ships rose. Within the broad ownership categories defined above, ships owned by governments and independents showed increases in both number of vessels and number of port calls. The number of oil company ships trading to the United States fell, although the number of port calls remained constant. The shift in ownership among vessels trading to the United States (more independents, fewer oil company) parallels the shift in ownership in the worldwide trading fleet. This was not the case for government-owned vessels, however. Government fleets in worldwide trade decreased between 1990 and 1994 (see Table 3-7), but government fleets trading to the United States increased by 35 vessels. This increase reflects a trend among oil producers—notably Saudi Arabia—to transport their crude oil to the United States in their own ships.

The total ownership figure for the U.S. trading fleet masks some significant data. The number of owners in the government and oil company categories is quite limited. The independent category is far more numerous and includes owners that have only one vessel as well as others whose fleets are in all respects comparable to those of the government or the oil companies.

As shown in Table 3-8, the Saudis increased their tonnage by more than nine times between 1990 and 1994, leaving all other state entities far behind. Kuwait increased its presence sixfold; China, fivefold. Ecuador, Uruguay, Finland,

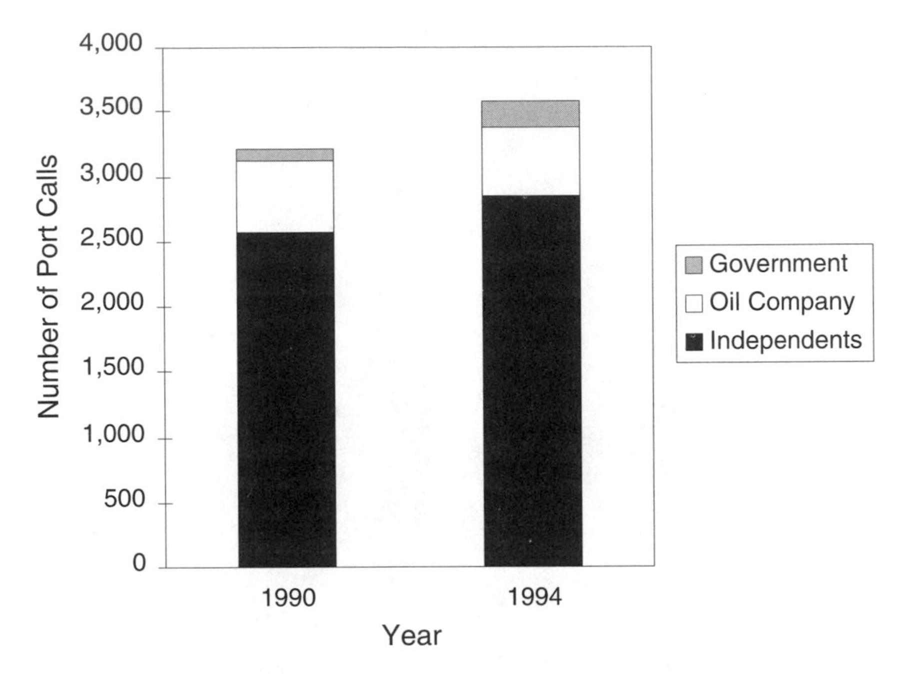

FIGURE 3-10 Changes in number of port calls, by ownership category, for U.S. trading fleet between 1990 and 1994.

TABLE 3-8 Tonnage of Government-Owned Fleets Trading to the United States, 1990 and 1994

Country	1990 Rank	1990 Tonnage (10^3 DWT)	1994 Rank	1994 Tonnage (10^3 DWT)
Saudi Arabia	1	1,176	1	10,881
Venezuela	2	1,106	5	1,503
Ecuador	3	891	—	0
China	4	805	2	4,152
Iraq	5	770	—	0
FSU	6	764	4	1,984
Spain	7	764	6	914
Kuwait	8	698	3	3,741
Brazil	9	250	—	0
Uruguay	10	87	—	0
Finland	11	82	—	0
Mexico	12	45	—	0
Egypt	13	0	7	873
Portugal	14	0	8	135

Note: FSU = former Soviet Union.

TABLE 3-9 Tonnage of Oil Company Fleets Trading to the United States, 1990 and 1994

Company	1990 Rank	1990 Tonnage (10^3 DWT)	1994 Rank	1994 Tonnage (10^3 DWT)
Chevron	1	24,893	1	31,647
Exxon	2	10,697	—	0
Texaco	3	8,754	4	4,048
Amoco	4	5,578	5	3,982
BP	5	5,379	7	810
Mobil	6	5,347	3	9,762
Shell	7	4,595	6	1,632
Total	8	3,640	8	781
Conoco	9	2,000	2	13,350
Coastal	10	956	—	0
Fina	11	853	—	0
Phillips	12	702	—	0
Hess	13	518	9	500 (estimate)[a]
Irving	14	263	11	114
O.K.	15	64	—	0
Apex	16	31	10	314

[a]Hess had replaced its tank ship fleet with integrated tank barges by 1994. Therefore data for Hess for 1994 were not included in data provided by Institute of Shipping Analysis, which were restricted to tank ships. The estimate of Hess' transport was provided by the company.

Mexico, and Iraq disappeared from the list, the latter as a direct result of the Gulf War. The disappearance of Mexico from the top 14 rankings in 1994 probably reflects the use of chartered vessels.

Dramatic changes also occurred in oil company fleets trading to the United States between 1990 and 1994 (see Table 3-9). Total oil company tonnage fell from 74.3 million DWT in 1990 to 66.4 million DWT in 1994. Over this period, Conoco increased its volume nearly seven times to move into second place among oil companies. Chevron and Mobil also increased substantially, while Hess stayed approximately the same. The U.S. trading fleets of all other companies either shrank or disappeared. Texaco, Shell, and Total reduced their tonnage by 50 percent or more. Exxon, Fina, Coastal, and Phillips gave up their fleets trading to the United States entirely, apparently shifting cargo movements from their own ships to chartered vessels.

When the amount of cargo discharged is analyzed in terms of the three U.S. geographic areas and the three ownership categories, it can be seen that the ownership distribution of discharging vessels changed significantly between 1990 and 1994 (see Table 3-10). The increase in government-owned vessels and tonnage on the Gulf and Atlantic coasts was due, in large part, to the increased presence of government-owned VLCCs in the Gulf and to increased crude oil

TABLE 3-10 Change in Ownership of U.S. Trading Fleet by Coast, 1990 and 1994

	Atlantic			Gulf			Pacific		
Ownership	1990 (10^6 DWT)	1994 (10^6 DWT)	Change (%)	1990 (10^6 DWT)	1994 (10^6 DWT)	Change (%)	1990 (10^6 DWT)	1994 (10^6 DWT)	Change (%)
Government	2.3	3.5	53	3.9	21.4	448	0.9	0.8	–11
Independents	87.5	88.6	1	152.2	195.9	29	8.9	14.5	63
Oil companies	19.7	10.9	–45	50.8	47.8	-6	3.8	7.7	103
Total	109.5	102.9	–6	206.9	265.2	28	13.6	23.0	69

imports from South America. The increased number of government-owned vessels trading to the Gulf coast came mainly from Saudi Arabia, Kuwait, and Venezuela (see also Table 3-8), reflecting a trend among oil producers to transport their own crude oil. The reduction in oil company activity on the Gulf and Atlantic coasts reflects their reduction in vessel ownership, particularly in trading to the United States.

The distribution of ownership on the Pacific coast did not change significantly in any sector. The Pacific trade is dominated by U.S. coastal trade from Alaska, and imports do not account for a major portion of Pacific traffic despite large percentage gains (21 percent in 1990 and about 31 percent in 1994).

JONES ACT FLEET

The U.S. coastal fleet of tankers consists of 116 vessels, 23 of which have double hulls. No new vessels were added to the fleet between 1990 and 1995, and 37 were scrapped (see Table 3-11). The committee was unable to establish

TABLE 3-11 U.S. Flag Vessels Sold or Scrapped, 1990–1995

Year	Scrapped	Sold	Total
1990	2	7	9
1991	4	2	6
1992	6	3	9
1993	10	3	13
1994	5	7	12
1995	10	9	19
Total	37	31	68

Sources: Clarkson Research, 1990–1995, 1996a,b.

whether any of the vessel sales were the result of OPA 90. There were no major changes in ownership or trading patterns during this period.

U.S. oceangoing barges play an important role in trade on both the Gulf and the East coasts but not the West coast. The total number of oceangoing barges is 93, of which 17 have double hulls. In addition, there are 13 integrated tug-barges, of which 3 have double hulls. Since the enactment of OPA 90, there have been only minor changes in the ownership and number of oceangoing barges and no flag changes.

Despite the apparent lack of influence of OPA 90 on the Jones Act fleet so far, Section 4115 is likely to have a strong impact in future years, as discussed in Chapter 5.

FINDINGS

Finding 1. Changes in vessel trading patterns to the United States in the 1990 to 1994 period were influenced primarily by changes in crude oil sourcing. The most notable change was a decrease in VLCCs carrying long-haul imports from the Middle East and a corresponding increase in vessels of 80,000 to 150,000 DWT carrying short- and medium-haul imports from Latin America and the Caribbean.

Finding 2. Some relatively new (i.e., built between 1985 and 1992) large, single-hull tankers are expected to trade in U.S. waters until 2015 for both regulatory and economic reasons. The deepwater port and lightering zone exemption of OPA 90 permits such vessels to continue in service to the United States through 2015. After 2010 there will be restrictions on these vessels discharging elsewhere in the United States, although they can continue to trade internationally until they are 30 years old. Long-haul crude oil imports are expected to increase from present levels, providing an economic impetus for the use of VLCCs that discharge their cargo through deepwater ports or lightering zones. Thus, the scrapping of a significant number of vessels of more than 150,000 DWT will be postponed, in part as a result of the OPA 90 exemption.

Finding 3. The proportion of double-hull tankers in the world fleet increased from approximately 4 percent in 1990 to 10 percent in 1994, reflecting the requirements of OPA 90 and IMO regulations.

Finding 4. Between 1990 and 1994, the average age of the fleet trading to the United States decreased by approximately one year, whereas the average age of the world fleet increased by one year. Vessels in the youngest (0 to 4 years) and oldest (20 to 24 years) age groups carried increased tonnage. These changes reflect both the construction boom of the mid-1970s and the relatively large number of recent double-hull deliveries.

Finding 5. The independently owned part of the world fleet continued to grow between 1990 and 1994, primarily at the expense of oil company vessel ownership; a similar trend was observed for the U.S. trading fleet. These changes in ownership patterns—some of which predate OPA 90—reflect a decision by some oil companies to leave the tanker business.

Finding 6. Government-owned fleets worldwide decreased between 1990 and 1994, whereas the percentage of government-owned tonnage trading to the United States increased from 2 percent to approximately 6.5 percent. This change largely reflects the growth in government-owned Saudi Arabian fleet tonnage trading to the United States. The sizes of the government-owned fleets of China and Kuwait operating to the United States also increased significantly.

REFERENCES

Clarkson Research, Ltd. 1990. Tanker Register. London: Clarkson Research.

Clarkson Research, Ltd. 1991. Tanker Register. London: Clarkson Research.

Clarkson Research, Ltd. 1992. Tanker Register. London: Clarkson Research.

Clarkson Research, Ltd. 1993. Tanker Register. London: Clarkson Research.

Clarkson Research, Ltd. 1994. Tanker Register. London: Clarkson Research.

Clarkson Research, Ltd. 1995. Tanker Register. London: Clarkson Research.

Clarkson Research, Ltd. 1996a. Shipping Review and Outlook: Spring 1996. London: Clarkson Research.

Clarkson Research, Ltd. 1996b. The Clarkson Register's Monthly Update: January 1996. London: Clarkson Research.

Drewry Shipping Consultants. 1994. The International Oil Tanker Market: Supply, Demand and Profitability to 2000. London: Drewry.

Drewry Shipping Consultants. 1996. The Shipbuilding Market Analysis and Forecast of World Shipbuilding Demand, 1995–2010. London: Drewry.

Energy Information Administration (EIA). 1994. Petroleum Supply Annual 1994. Washington, D.C.: U.S. Department of Energy.

Energy Information Administration (EIA). 1996. Annual Energy Outlook 1996. Washington, D.C.: U.S. Department of Energy.

Institute of Shipping Analysis. 1996. Study prepared for the Committee on Oil Pollution Act of 1990 (Section 4115) Implementation, Göteborg, Sweden.

Jacobs & Partners Limited. World Tanker Review: July–December 1995. London: Jacobs & Partners Limited.

Navigistics Consulting. 1996. Tanker Supply and Demand Analysis, Task 1 and Task 2. Study prepared for the Committee on Oil Pollution Act of 1990 (Section 4115) Implementation Review. Washington, D.C., June.

Petroleum Industry Research Foundation, Inc. (PIRA). 1992. Transporting U.S. Oil Imports: The Impact of Oil Spill Legislation on the Tanker Market. Prepared for the U.S. Department of Energy. New York: Petroleum Industry Research Foundation, Inc.

Tanker Advisory Center, Inc. 1991. 1991 Guide to the Selection of Tankers. New York: Tanker Advisory Center.

Tanker Advisory Center, Inc. 1995. 1995 Guide to the Selection of Tankers. New York: Tanker Advisory Center.

Tanker Advisory Center, Inc. 1996. 1996 Guide to the Selection of Tankers. New York: Tanker Advisory Center.

Wilson, Gillette & Company. 1994. The Evaluation of Past and Future Crude Oil Lightering Operations in the U.S. Gulf Coast. Arlington, Va.:Wilson, Gillette & Company.

4

Economic Impact of the Oil Pollution Act of 1990 on the International Tanker Fleet

The purpose of this chapter is to assess the economic impact of the Oil Pollution Act of 1990 (OPA 90), Section 4115 and MARPOL Regulations I/13F and I/13G (MARPOL 13F and 13G) on the operations of the international tanker fleet. The committee concluded that the likely economic impact of these regulations will consist mainly of increased costs arising from (1) increased capital expenditures to replace retired tankers with double-hull tankers; (2) increased operating expenses of double-hull tankers; and (3) costs associated with the timing of the replacement of single-hull vessels.

The replacement of existing tankers with double-hull instead of single-hull tankers will be the costliest of these items. As existing tankers are phased out, supply will eventually fall below actual (and perceived) demand. The resulting gap between supply and demand will be filled by the construction of new tankers or the conversion of existing ones. Beginning January 1, 1994, new and converted tankers must have double hulls. The committee has used two approaches toward assessing the evolution of tanker demand and supply to ascertain the likely characteristics of the new double-hull tanker fleet: (1) a computer model using certain assumptions to isolate the impacts of Section 4115 and MARPOL 13F and 13G and (2) an analysis of how "real-world" events may differ from the computer model. The international market is characterized by the highly competitive structure and conduct of the participants.[1] Accordingly, demand, supply,

[1]The competitive structure arises from: (1) the large size of the tanker market with units owned or controlled by independent owners and ship users (oil companies and oil producers); (2) very low concentration of buyers and sellers with the largest controlling less than 5 percent of total supply; (3) tanker firms widely distributed geographically; (4) tankers that deliver an essentially homogeneous service with little differentiation; (5) low barriers to entry; and (6) a high degree of efficiency and availability of expert knowledge.

new construction, and scrapping are determined mainly by freight rates, although investment in new tankers may be driven by market sentiment rather than by strict economics.

TANKER SUPPLY AND DEMAND

At year-end 1995, the world tanker fleet consisted of about 3,000 tankers with a total capacity of 255 million deadweight tons (DWT). In addition, combined carriers designed to carry ore, dry bulk, or oil as cargo, totaling 25 million DWT, raised the potential capacity to 280 million DWT.[2] To determine the size and makeup of the new double-hull tanker fleet arising from the implementation of OPA 90 and MARPOL, it was necessary to determine the number of tankers that will be needed to carry oil in the future, as well as the current and prospective supply of tankers. Estimates of future tanker demand should include both the new construction necessary to replace the existing fleet as it ages or is forced to retire and the new construction needed to carry any increase in demand for oil.

Tanker Supply

The differing impacts of Section 4115 and MARPOL 13F and 13G on the wide range of vessels used in the maritime oil transportation industry required the committee to develop a methodology for determining the effective supply of tankers over the next 20 years. (The methodology, developed in conjunction with industry experts, is defined in Appendix F.)

Vessels in the international fleet were assigned to one of four categories: double-hull vessels, double-side or double-bottom vessels, and single-hull vessels, either pre-MARPOL or MARPOL (see Figure 4-1).

Because both Section 4115 and 13F and 13G use vessel age as a criterion, the age profile of the fleet will be a significant determinant of how these regulations change the composition of the fleet. The age of the current fleet continues to reflect the impact of the shipbuilding boom of the mid-1970s. Peak capacity is found in vessels between 16 and 20 years of age, as can be seen in Figure 4-2.

The future impact of Section 4115 on the size of the international tanker fleet able to call at U.S. ports is shown in Figure 4-3. The sharp decline in 2015 marks the end of the deepwater port and lightering zone exemption that allows single-hull tankers (assumed to be tankers exceeding 120,000 DWT) to continue trading to the United States until that year.

Section 4115 does not force tankers to retire but bans them from trade to the United States. Tanker retirement, however, will be forced by the phaseout

[2]The role of combined carriers in the oil transportation market is declining in importance. Combined carriers now account for about 14.5 percent of carrying capacity compared with 25 percent in the early 1980s; this figure is expected to fall below 7 percent by 2005.

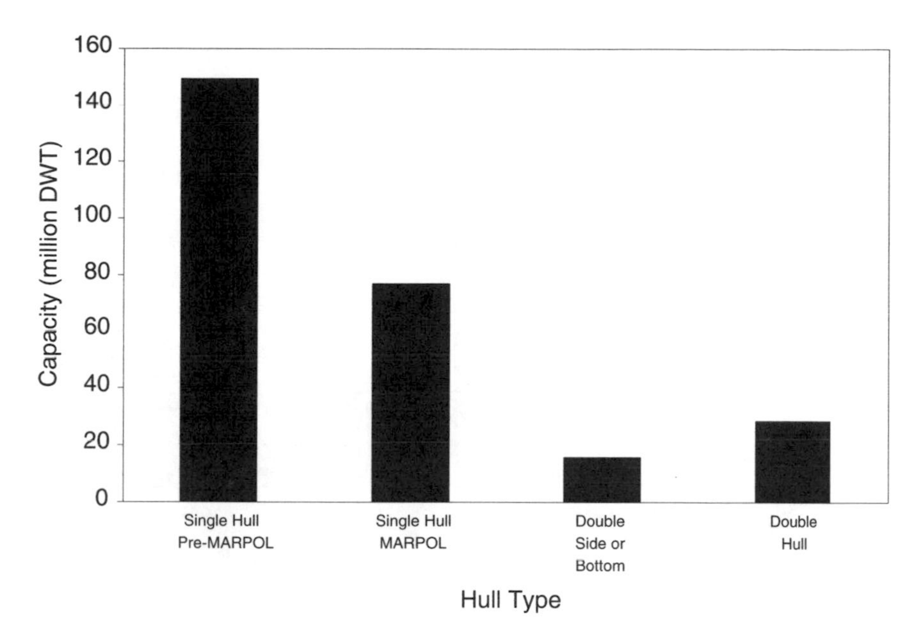

FIGURE 4-1 Capacity of international tanker fleet by hull type as of October 1995. Source: Navigistics, 1996.

schedule of 13G.[3] The retirement by year of international tankers as a consequence of 13G is shown in Figure 4-4, where it is compared to the impact of Section 4115.[4] Regulation 13G will have the greatest impact during the period 2004 to 2008, when vessels constructed during the shipbuilding boom of the mid-1970s will reach the 30-year age limit. To the extent that 13G affects trading to the United States, some tankers will be retired earlier than they might have been under the OPA 90 lightering zone exemption. The figure does show, however, that Section 4115 will have a stronger impact because of its 25-year age limit, as opposed to the 30-year limit of 13G.

Figure 4-4 assumes that pre-MARPOL tankers will continue trading to 30 years using hydrostatically balanced loading (HBL), rather than being retired at 25 years—in other words, HBL will extend the retirement age of tankers built during the boom of the mid-1970s by five years, beginning in 1999. Thus, a tanker that reaches age 25 in 1999 is assumed to operate until 2004. This impact will largely be eliminated by 2008. Figure 4-5 shows the impact on the

[3]For this analysis the committee has assumed that MARPOL 13G is ratified by all countries except the United States.

[4]It is assumed that all non-double-hull vessels will continue trading until they reach 30 years of age. Pre-MARPOL tankers will hydrostatically load (with an assumed average loss in cargo capacity of 8 percent of deadweight).

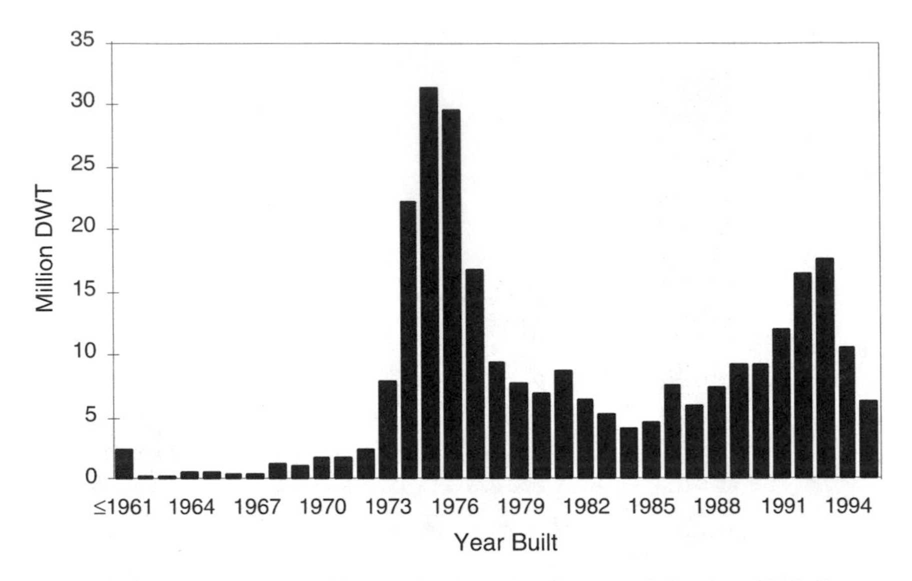

FIGURE 4-2 Age profile of international tanker fleet as of October 1995. Source: Navigistics, 1996.

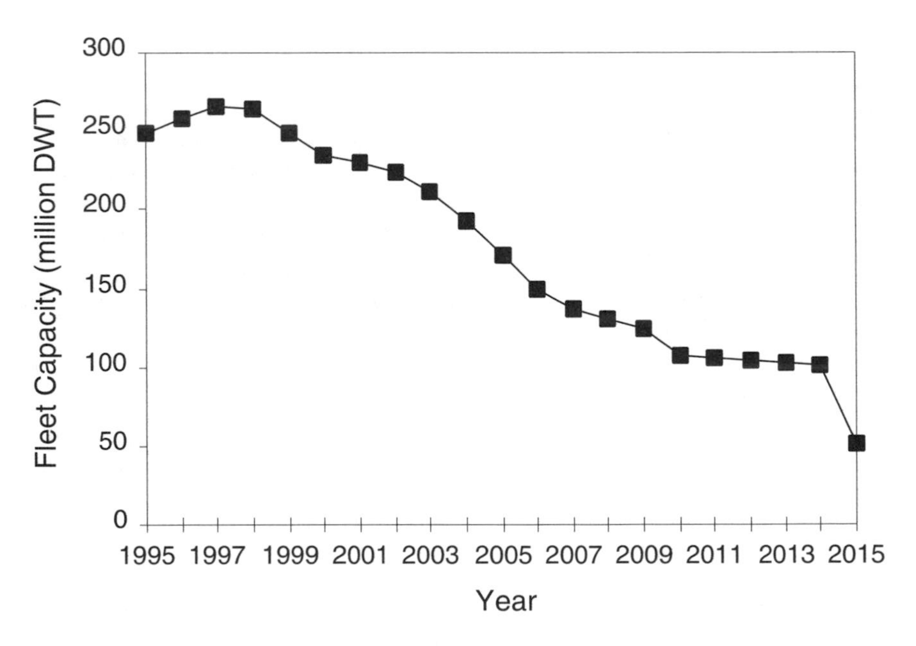

FIGURE 4-3 Impact of OPA 90, Section 4115, on size of international tanker fleet eligible to trade in U.S. waters. Source: Navigistics, 1996.

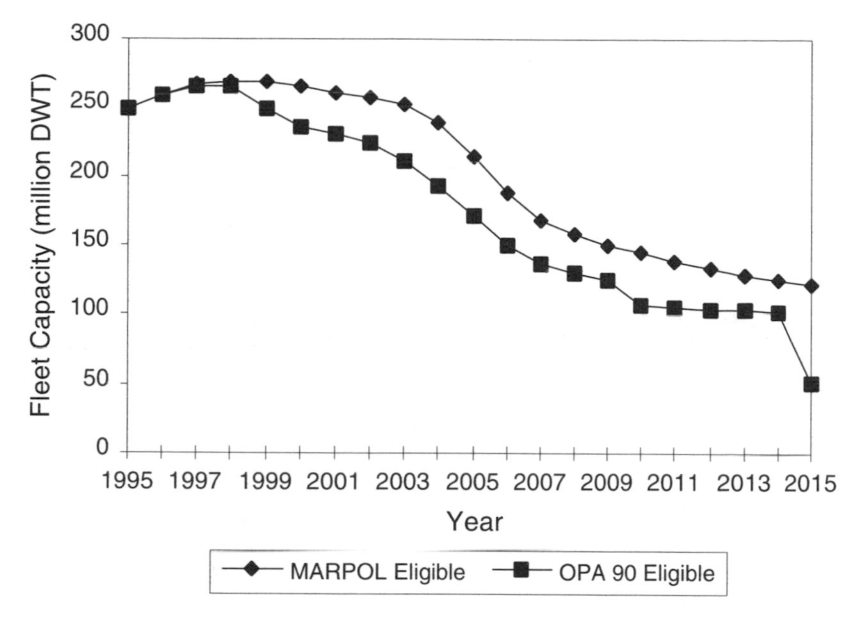

FIGURE 4-4 Impact of MARPOL 13G on the size of the international tanker fleet. Source: Navigistics, 1996.

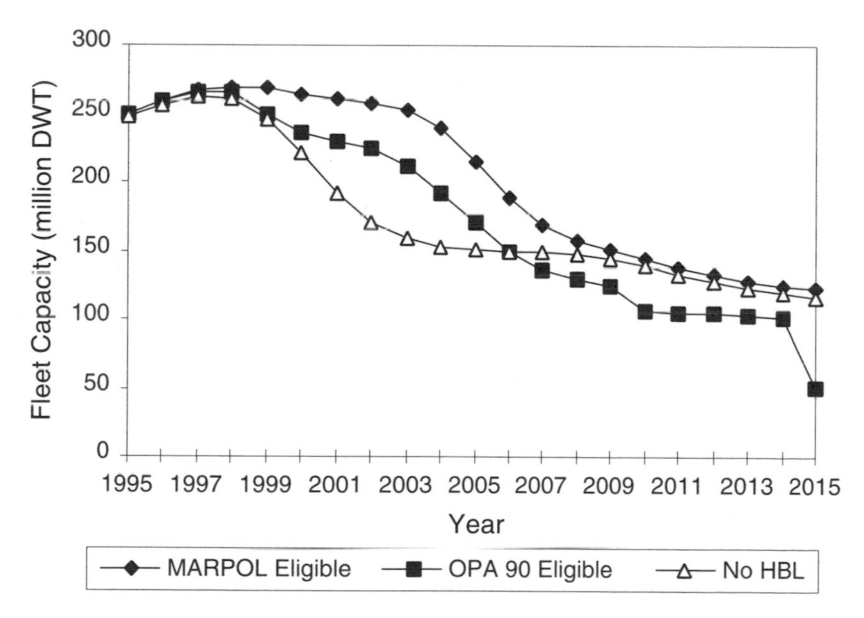

FIGURE 4-5 Impact of HBL alternative on the size of the international tanker fleet through 2015. Source: Navigistics, 1996.

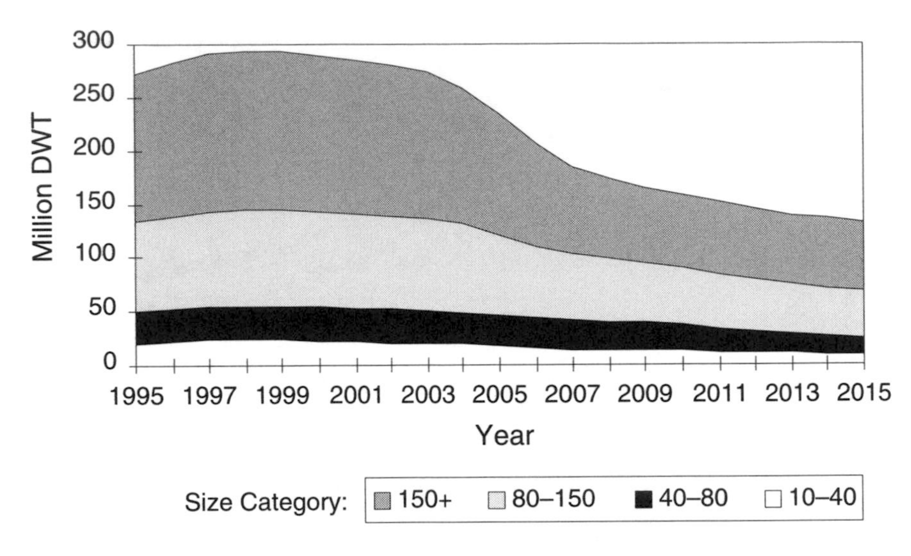

FIGURE 4-6 Impact of MARPOL 13G on the international tanker fleet by size category through 2015.

international fleet if no vessels utilize HBL and compares this result to the impacts if all vessels implement HBL. Figure 4-6 shows the projected fleet, as affected by 13G, by size category (including orders for new construction as of October 1995) from 1995 to 2015.

International Tanker Demand

The analysis of demand for tankers is restricted to the 1995 to 2005 period because the committee determined that estimates of oil demand after 2005 are too unreliable for coherent analysis. Projections of oil demand were provided by PIRA Energy Group (1995) and compared with similar projections by Marine Strategies International (1996a) and the U.S. Department of Energy (EIA, 1996). Figure 4-7 shows projected crude and petroleum products oil flows to 2005. Crude oil movements, which account for about 80 percent of the tanker trades, grow at an annual rate of 3.3 percent, whereas oil products continue relatively flat.

The projected pattern depicted in Figure 4-7 masks important changes in the underlying trades that will have a considerable impact on tanker markets (PIRA, 1996). In the U.S. trade, short-haul imports from Latin America are projected to double, offsetting a decline in long-haul shipments from the Middle East and Africa. In Western Europe, crude imports decline by 10 percent between 1995 and 2000, followed by 38 percent growth to 2005. Imports of Middle East crude to Western Europe decline by 16 percent by 2000 but reverse with a 44 percent increase by 2005. The projected declines between 1995 and 2000 are offset by sharply increasing crude imports from the former Soviet Union (FSU), which more than double by 2005. China, Southeast Asia, and Japan continue as markets of rapid growth through 2005, accounting for more than 60 percent (compared to 40 percent currently) of Middle East crude exports and resulting in a substantial reduction in longer-haul shipments. The shift from long to shorter hauls—notably in trade to the United States, China, Southeast Asia, and Japan—indicates that for the next few years, the increasing oil flows traced by Figure 4-7 will not be translated into corresponding increases in tanker demand because individual tankers traveling over shorter distances will be able to carry more oil per year.

Although the committee relied on the PIRA base case forecast to calculate tanker requirements, there are other credible scenarios of tanker demand. Hence, the future demand for tankers cannot be predicted with confidence (Stopford, 1996). For instance, PIRA assumes an early resumption of Iraqi oil exports that will displace 7 million DWT of very large crude carriers (VLCCs) and a rapid rate of FSU exports that will displace another 10 million DWT because both Iraqi and FSU oil exports will utilize pipelines for all or part of their transportation. If any of these assumptions fail to materialize, tanker demand will be significantly altered; the examples cited would change tanker demand by 17 million DWT. In the committee's judgment, the PIRA analysis is the most realistic.

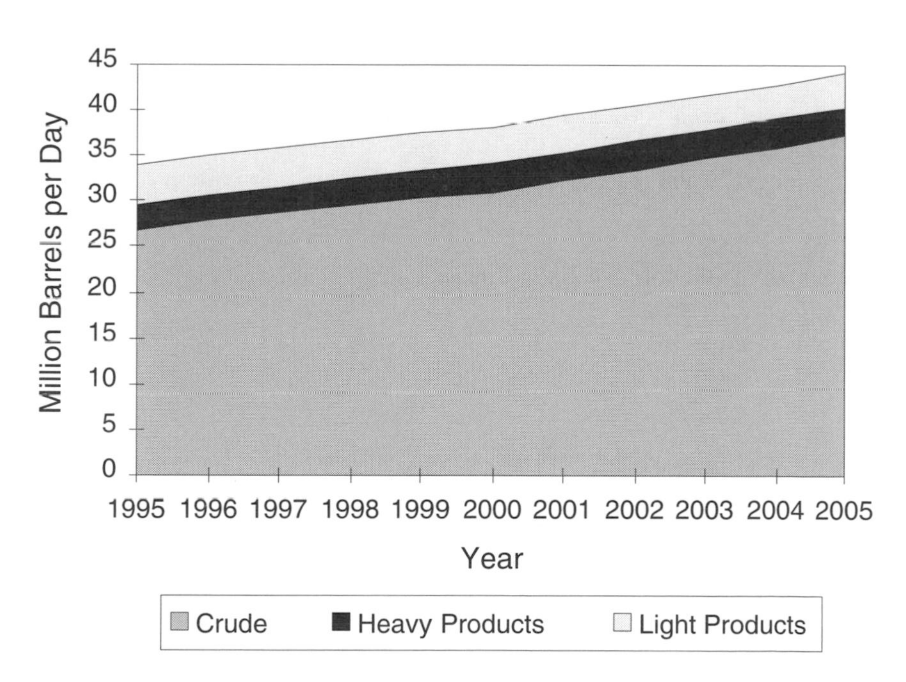

FIGURE 4-7 International tanker oil flows, 1995–2005. Source: PIRA, 1995.

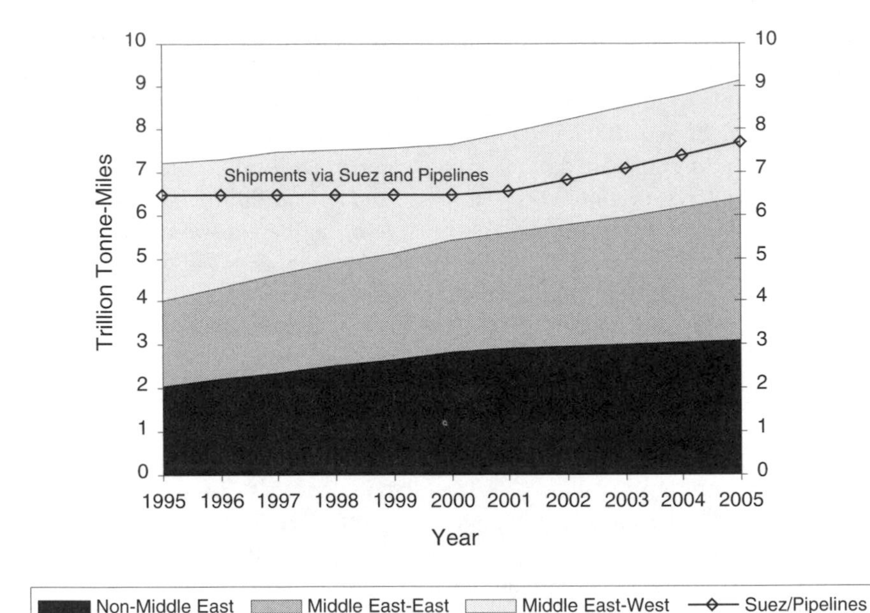

FIGURE 4-8 Interregional crude oil exports by region, 1995–2005. Source: PIRA, 1995.

Crude oil exports by region of origin between 1995 and 2005 are summarized in Figure 4-8. This figure illustrates the relative importance of Middle East exports of crude and the vigorous pace of such exports to eastern destinations. Also shown is the impact attributed to pipelines and the Suez Canal, which shorten voyage distances and reduce tanker requirements. The extent and timing of any resumption of Iraqi oil exports will have a significant impact on near-term demand. Figure 4-9 translates projected oil flows into tanker requirements for the transport of crude oil and products. Tanker demand in the projection remains stagnant to the end of the 1990s and resumes growth in the next century.

SUPPLY-DEMAND BALANCE

Impact of HBL Requirements for New Construction

Tanker supply is defined here as the existing fleet of international tankers and combined carriers, augmented only by newbuildings on order as of October 1995 and diminished only by the mandatory phaseout provisions of OPA 90 and MARPOL. This definition is designed to isolate the impacts of Section 4115 and 13G.

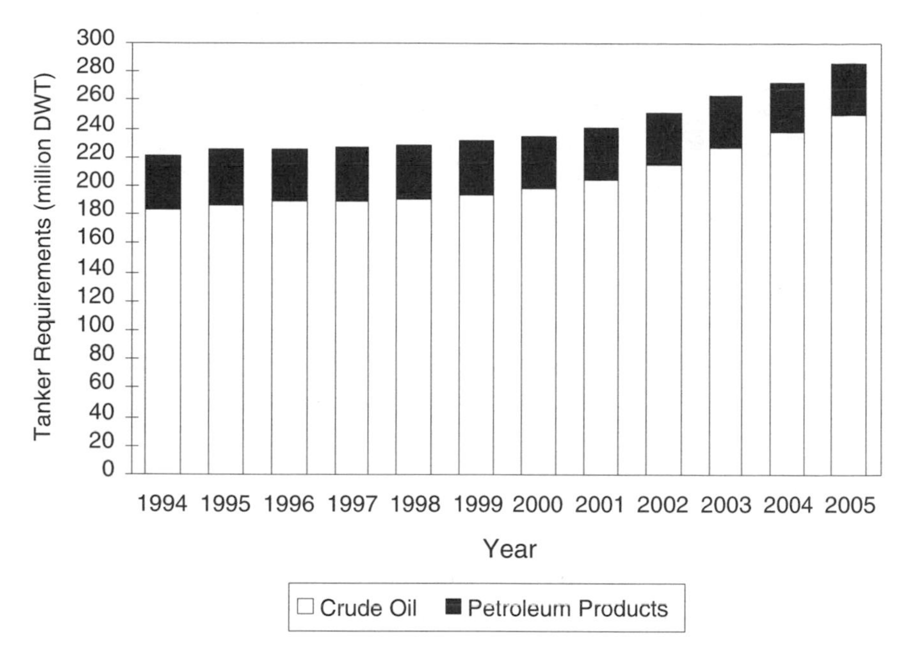

FIGURE 4-9 Tanker requirements for the transportation of crude oil and petroleum products, 1994–2005. Source: PIRA, 1995.

Under competitive conditions, when current demand approaches current supply, freight rates rise (see Appendix G). If future freight rates are perceived as generating a profit, newbuildings are ordered. The operation of a competitive market, therefore, is self-adjusting in that any deficiency in tanker supply induces the construction of new tonnage.

By subtracting demand from potential supply, the tanker surplus (or deficiency) can be derived and should be equal to the minimum tanker capacity (ordered after October 1995) necessary to offset the deficiency. In the following analysis, three scenarios are used to estimate newbuilding requirements: owners take delivery of new vessels (1) only when demand equals 100 percent of maximum supply, including HBL and all oil-bulk-ore vessels (OBOs); (2) when demand reaches 95 percent of maximum supply; and (3) when demand reaches 95 percent of supply without HBL. The results are shown in Figure 4-10.

The first two scenarios are designed to isolate the impacts of OPA 90 and MARPOL. The first scenario estimates that newbuilding deliveries start in 2004 and total about 55 million DWT by 2005. The second scenario estimates that newbuilding deliveries start in 2003 and will amount to about 66 million DWT by 2005. Under the third scenario, retention of existing capacity does not occur; newbuilding requirements begin in 2000; and by 2005, new deliveries total about

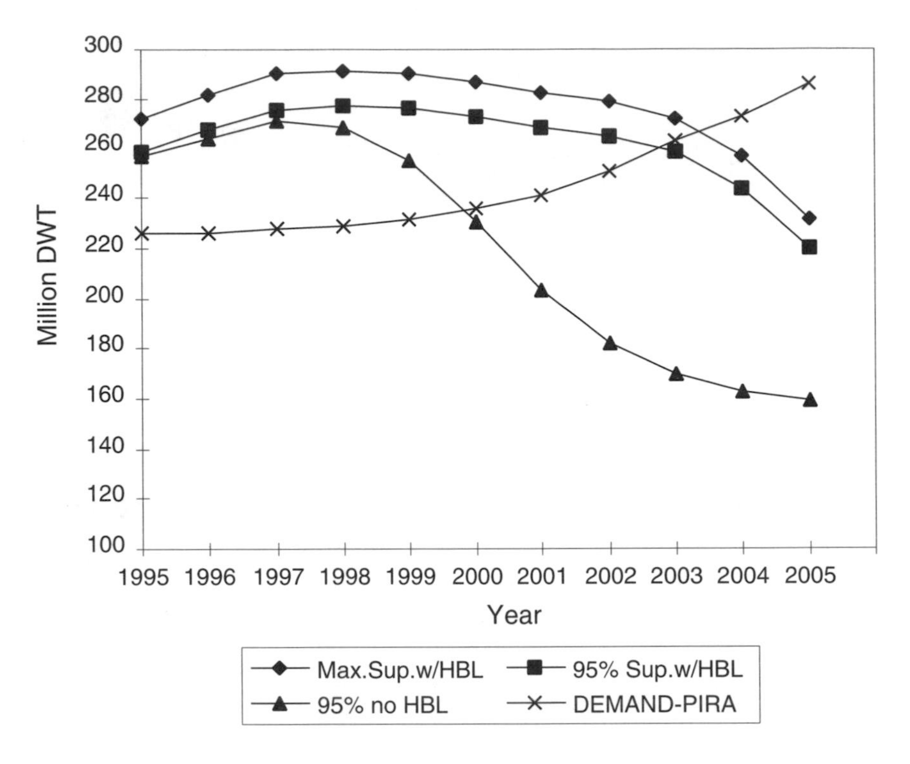

FIGURE 4-10 Aggregate supply-demand tanker balance with and without HBL, 1995–2005. Note: Max = maximum; Sup = supply.

127 million DWT. The newbuilding requirements under the three scenarios are summarized in Figure 4-11.

Hence, the key item in determining the required level of newbuildings will be the extent to which the HBL option is utilized by shipowners during the next few years. Clearly, the HBL alternative, if fully utilized, would provide a huge amount of equivalent tonnage. At its peak in 2003, the HBL option provides close to 90 million DWT, or one-third of the current tanker fleet (Table 4-1).

The extent to which shipowners adopt HBL will be determined by the condition of each tanker, the costs of life extension, and prevailing freight rates (MSI, 1996a). So far, no VLCCs have operated beyond age 25, and the average age of VLCCs scrapped in the last five years was much lower (see MSI, 1996a, and Figures 3-3 and 3-4). Some owners may forgo the HBL alternative and scrap their tankers earlier to avoid paying the high price of the special surveys required for a shortened period of service in a market increasingly favoring modern tankers. The self-adjusting market mechanism notwithstanding, the unsynchronized decisions of owners planning newbuildings and those facing scrapping of old ships

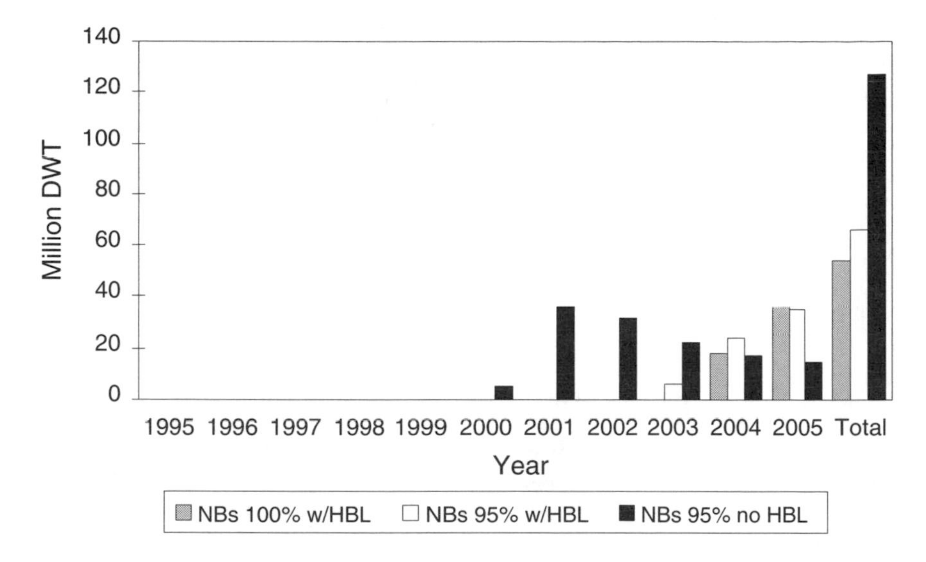

FIGURE 4-11 Tanker newbuildings required under MARPOL 13G for 1995–2005 with and without HBL. Note: NBs = newbuildings; w = with.

(who may be different people) could lead to periods of temporary imbalance with a significant short-term impact on freight rates. Moreover, seasonal spurts in demand (see Appendix H) could temporarily put pressure on capacity and cause freight rates to rise, inducing some shipowners to place early orders for newbuildings.

For these reasons, analysts are predicting a gradual tightening of tanker tonnage balances and increasing orders for new tankers in the next few years. Both

TABLE 4-1 Additional Fleet Capacity in Million DWT after Adoption of HBL

Year	Additional DWT
1995	1.6
1996	3.7
1997	4.9
1998	9.0
1999	21.2
2000	41.7
2001	65.1
2002	82.8
2003	88.8
2004	81.3
2005	61.0

Marine Strategies International (MSI) and Clarkson predict that double-hull tanker deliveries will average 20 million DWT annually between 1998 and 2000 and continue at a high level through 2004 (Stopford, 1996). PIRA forecasts that by the end of the 1990s, "tanker orderbooks are likely to be at the highest level in over 20 years" (PIRA, 1996).

Regardless of the timing of deliveries of new double-hull vessels, it is very likely that the market will tighten over the next two to three years because of the aging of the many tankers built in the 1970s, an accelerated rate of scrapping induced by OPA and MARPOL, and the current low level of tanker orders, which is the lowest in eight years.

Supply-Demand Balance by Vessel Size

Projections of demand for the international tanker fleet as a whole are shown in Figure 4-12; projections of supply and demand by vessel size are shown in Figure 4-13. In the early years a tighter balance is indicated for smaller tankers, whereas a significant longer-lived surplus is suggested for the larger ships, particularly VLCCs. In Figure 4-13, the apparent shortage in tanker capacity from

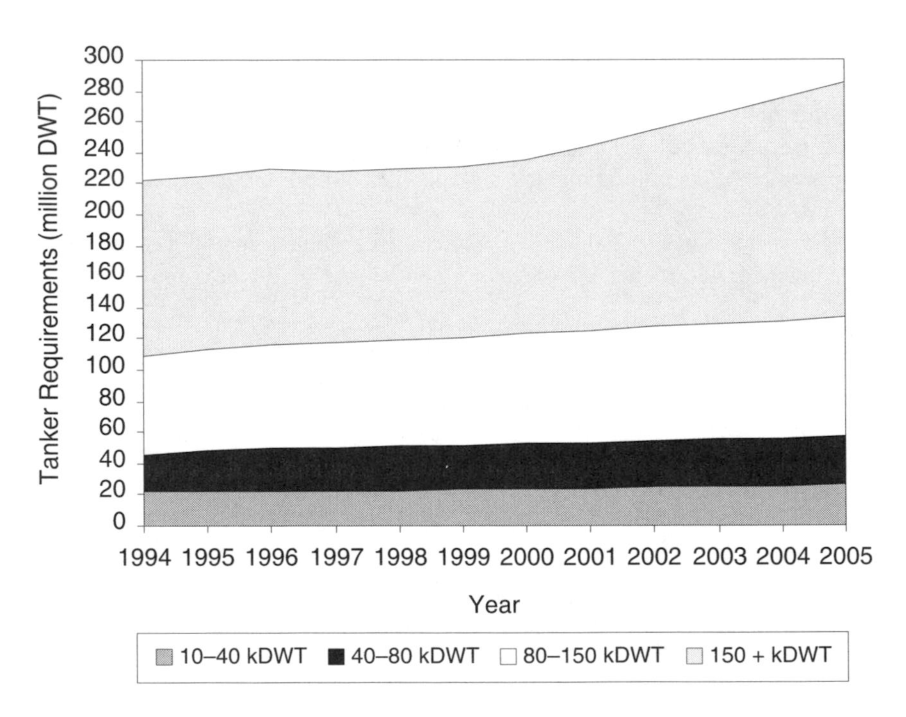

FIGURE 4-12 International tanker requirements for all size segments, 1995–2005. Note: kDWT = 10^3 DWT.

1995 to 1998 for the 10,000 to 40,000 DWT segment arises from the exclusion of the sizable fleet of chemical tankers, many of which also carry oil.

Requirements for larger tankers of more than 150,000 DWT shown in Figure 4-13 are based on the different forecasts of PIRA and MSI. Because this segment is the most vulnerable to HBL penetration, it tends to dominate the supply-demand balance in the entire market. Additionally, the deepwater port and lightering exemption of OPA 90 will enhance the appeal of using HBL for larger vessels.

Factors Influencing Supply and Demand

The previous discussion suggests an orderly pattern of scrappings and new construction determined by the interaction of supply and demand, albeit with uncertainty regarding the use of HBL. However, a strictly rational approach must be modified by other factors, including freight rates that are at this time substantially lower than those required to support newbuilding, the apparent absence of a two-tier market that would reward double-hull tankers, the continuing change in the tanker market structure, and the present dearth of orders for new construction.

Required Freight Rate

The charts in Figure 4-14 compare the required freight rate (RFR) of typical tankers with actual rates in 1992 through 1995 and the first half of 1996. RFR is defined as the rate required to cover tanker operating expenses and realize a desired return on capital, assumed in this case to be 10 percent. The charts show that actual freight rates during the period averaged around 60 percent of the RFR, with VLCCs averaging about 57 percent, Suezmax tankers 65 percent, and the others in between. Actual rates improved during the first half of 1996 but were still less than 70 percent of the RFR for all vessel classes.

The overall picture reflects the depressed state of the international tanker market from 1992 to 1995. Average market rates for a one-year time charter for modern tankers have been less than two-thirds of the RFR. The depressed freight rates of 1992 to 1995 are typical of the last 20 years. For 16 out of those 20 years, modern tankers have been chartered below cost (Stopford, 1996). Overall, the return on net assets in the tanker business during 1990 to 1994 period was about 1.4 percent for product carriers, 3.8 percent for Aframax tankers, and zero for tankers of more than 100,000 DWT. Many tankers showed a negative return. Freight rates are now higher, but they must rise even further if they are to induce the construction of enough double-hull tankers to replace retiring single-hull ships, unless shipbuilding capacity is so high that tankers can be delivered below cost, which seems unlikely.

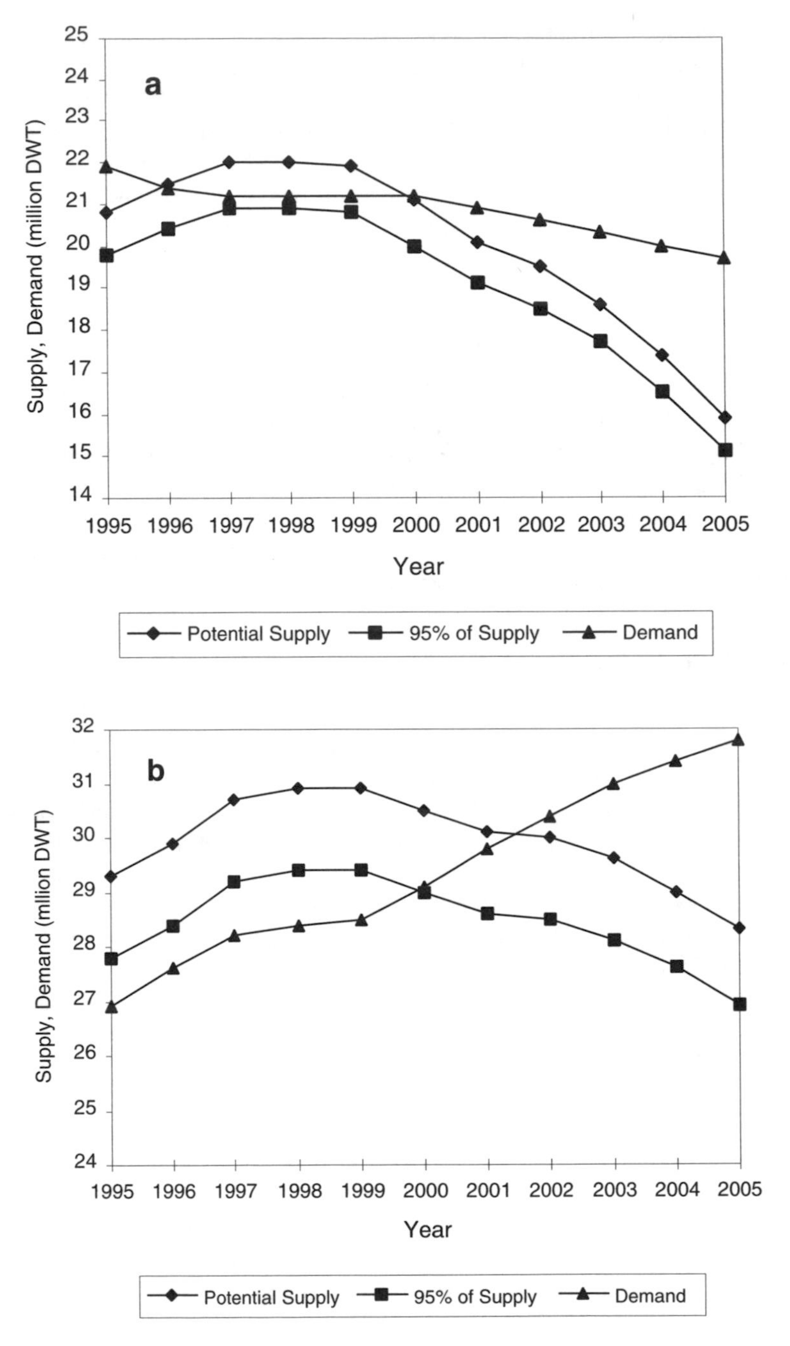

FIGURE 4-13 International tanker requirements for individual size segments, 1995–2005. (a) 10–40 kDWT (10^3 DWT); (b) 40–80 kDWT; (c) 80–150 kDWT; (d) 150 + kDWT.

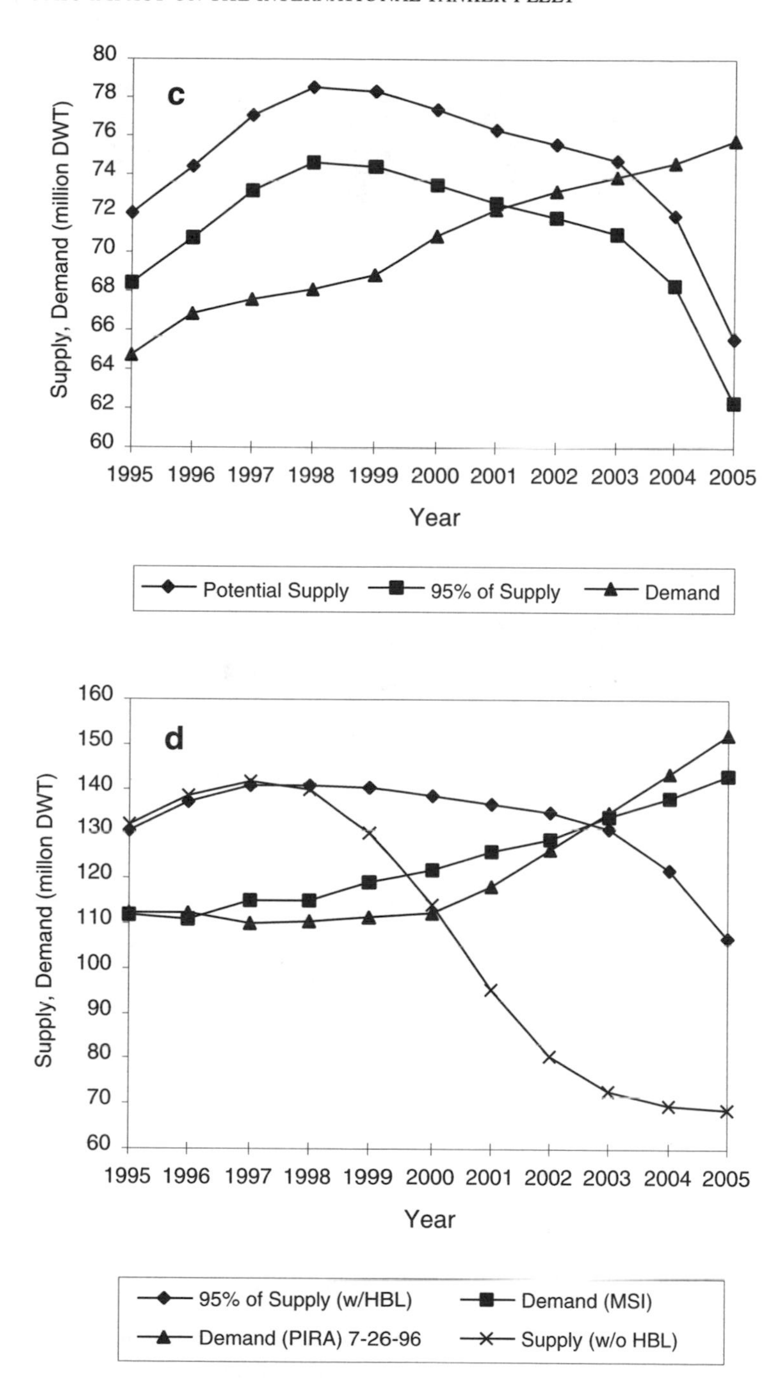

FIGURE 4-13 *Continued*

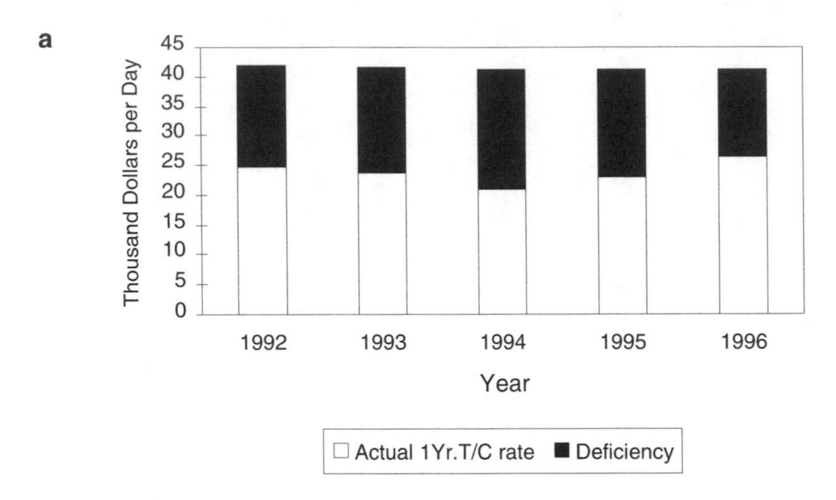

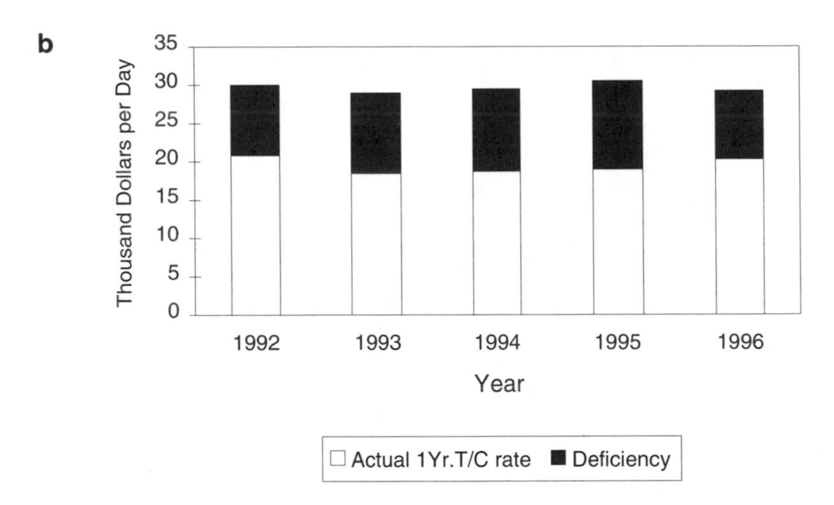

FIGURE 4-14 RFR by market segment, 1992–1996. Data for first six months of 1996: (a) VLCCs; (b) Suezmax tankers; (c) Aframax tankers; (d) handy-size product tankers. Note: T/C = time charter. Source: MSI, Freight Forecaster, 1996b.

Two-Tier Freight Rate Market

Shortly after enactment of OPA 90, it was anticipated that modern tankers in general, and double-hull tankers in particular, would charge a higher freight rate to compensate for their higher quality and to help cover their higher cost of construction. The committee has obtained information from several experts and has reviewed available data to determine if any such rate premium was paid for newer

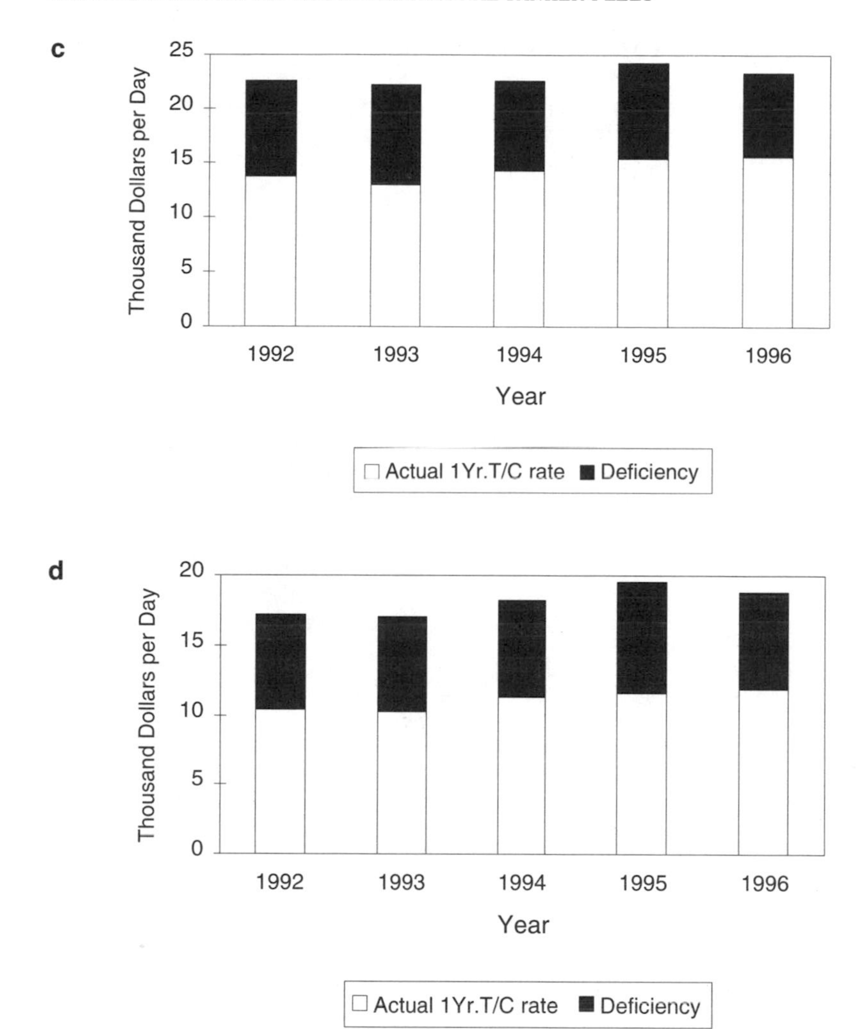

FIGURE 4-14 *Continued*

ships, for double-hull construction, or for trading to the United States. In general, ship brokers and other experts declined to affirm the existence of a rate differential for tankers in these categories (Jones, 1995; Loucas, 1995; Shawyer, 1995). They did believe, however, that given equal rates, quality tankers would be preferred. These views were summarized in a letter to the committee from the London Tanker Brokers' Panel (see Appendix I).

Others present a different view. A rigorous statistical analysis has been

TABLE 4-2 Two-Tier Markets after OPA 90

Vessel Size Category	General Comparisons	Specific Comparisons
60,000–80,000 DWT	U.S.-bound tankers received a higher rate than non-U.S.-bound tankers	Single-hull U.S.-bound tankers received a premium averaging 15.6 to 19.6 percent more than single-hull non-U.S.-bound tankers. Old (more than 10 years) U.S.-bound tankers received a premium averaging 13.8 percent more than old non-U.S.-bound tankers
80,000–50,000 DWT	U.S.-bound tankers received a premium ranging from 6 to 12 percent more than non-U.S.-bound tankers	Modern (less than 10 years old) double-hull U.S.-bound tankers received a higher (12.3 percent) premium than modern, double-hull, non-U.S.-bound tankers. Statistical analysis focusing on age alone failed to establish a consistent pattern
150,000 DWT or more	Double-hull tankers, mostly VLCCs and ULCCs, received a premium averaging WS 18.5 (36 percent) for U.S.-bound tankers and WS 8.1 (16 percent) for non-U.S.-bound tankers	In this size group, the number of observations was limited

Note: ULCC = ultralarge crude carrier; WS = Worldscale.

performed to determine whether a quantifiable premium was paid for modern, double-hull tankers in general and for tankers trading to the United States in particular (Tamvakis, 1995). The study examined 14,000 crude oil spot fixtures covering 1,600 crude oil tankers of 60,000 DWT or more from January 1, 1989, to June 30, 1993. The data were grouped into tanker size segments and time periods before and after the enactment of OPA 90. With respect to the post-OPA 90 period, tankers bound for the United States (regardless of size) received higher rates than tankers bound for other destinations (see Table 4-2).

The Tamvakis (1995) study may chiefly reflect the impact on U.S. rates of the liability features of OPA 90 rather than its double-hull requirement. The greater risks and costs embodied in guarantees of financial responsibility and higher protection and indemnity (P&I) premiums[5] have, in effect, imposed a

[5]The Worldscale system has included a fixed rate differential for tankers trading to the United States to cover additional P&I premiums since July 1, 1993.

penalty on shippers delivering cargo to the United States. In non-U.S. trade, VLCCs were the only group of double-hull tankers that appeared—on the basis of limited statistical evidence—to earn premium rates. This finding may reflect the efforts of some nations, notably Japan and Finland, to attract vessels of superior design and below-average age. In any event, the differential rate that has prevailed falls short of what is needed to compensate for the capital cost of new construction.

Change in Market Structure

In the last 20 years, the market for tankers has witnessed a gradual but significant change. As the cost of oil transport fell to a small fraction of the delivered price, some oil companies began to retreat from the tanker business. Increasingly, the responsibility for transport has shifted from oil companies to oil traders (Stopford, 1996).

Although the size of the tanker fleet as a whole increased by more than 50 percent between 1973 and 1994, seven major oil companies[6] cut their company fleets in half, a decline from 43 million DWT (599 ships) to 22 million DWT (184 ships). This represented a 49 percent reduction in tonnage and a 69 percent reduction in ships. The oil companies' share of the international fleet trading to the United States fell from 23 percent of the tonnage employed in 1990 to 17 percent in 1994.

Chartering has also changed considerably. The proportion of the independent fleet on time charter (long-term contracts) to oil companies decreased from 80 percent in 1973 to 20 percent in 1988, whereas independent tankers hired on short-term contracts increased from 20 to 80 percent over the same period. In 1995, 2,098 tankers completed 14,409 spot market fixtures (Stopford, 1996).

Reduction in Tanker Construction Orders.

Prevailing freight rates and changes in market structure have contributed to a significant reduction in orders for new tankers. Construction orders as of July 1, 1996, are shown in Table 4-3. Only about 16 million DWT of crude and product tankers were on order in June 1996, and only 27 crude and product tankers, amounting to 3.2 million DWT (2.2 percent of the existing fleet on an annualized basis), were ordered during the first half of 1996. In 1991, orders reached 6 percent of the tanker fleet. In 1997, scheduled deliveries will fall to 2 percent of the tanker fleet (Clarkson, 1996a).

This decline is not attributable to OPA and MARPOL, because mandatory phaseouts would ordinarily have encouraged new construction. The sluggishness in orders for new tankers is apparently the result of an oil transportation market

[6]Amoco, British Petroleum, Chevron, Exxon, Mobil, Shell, and Texaco.

TABLE 4-3 Tanker Fleet and Orderbook as of July 1, 1996

Vessel	Segment (10^3 DWT)	Fleet (10^6 DWT)	Orderbook (10^6 DWT)			
			1996	1997	1998+	Total
VLCC	200+	125.3	4.0	1.8	1.2	7.0
Suezmax	120–200	39.8	1.1	1.7	0.3	3.1
Aframax	80–120	41.7	1.0	2.0	1.3	4.3
Panamax	60–80	14.6	0.1	0.0	0.0	0.1
Small	10–60	40.5	1.5	0.2	0.0	1.7
Total		261.9	7.7	5.7	2.8	16.2

Source: Clarkson Research, 1996a.

that is strong enough to be profitable for older single-hull tankers (hence, a low scrapping rate) but not strong enough to pay the costs of new double-hull tankers (PIRA, 1996). Considering the higher freight rate required to cover the cost of a new tanker, the absence of an adequate two-tier market, and the dominance of spot market fixtures over period charters, independent owners (who hold 65 percent of the fleet) are reluctant to order new tonnage. Despite the certainty of mandatory scrapping of vessels and favorable shipbuilding prices, tanker orders are currently at their lowest level in more than eight years.

If tanker orders do not resume, the market is likely to enter a temporary period of sharp freight rate volatility. Pressures to build would then rise, and a rush to build might cause a temporary glut of orders, but shipbuilding capacity appears adequate to accommodate any increased level of construction. The market will discourage any pronounced long-term bunching of orders. If freight rates are low, resulting in few orders, the price of construction will go down so that shipyards can be better utilized. Alternatively, if freight rates increase and many orders are placed, prices will rise, thereby discouraging further orders.

ADEQUACY OF WORLD SHIPBUILDING CAPACITY AND FINANCING

Estimates of Shipbuilding Capacity and Demand

Recent reports from Europe and Japan conclude that worldwide shipbuilding capacity over the next several years will exceed demand. A European report (de Albornoz, 1996) states that capacity fell during the 1980s to a low of approximately 15 million compensated gross tons (CGT)[7] by 1990 but that it has been

[7]Compensated gross tons is a term used to describe the capacity of a shipyard that reflects the complexity of vessel construction—a cruise vessel, for example, will require more cost and time than a tank ship.

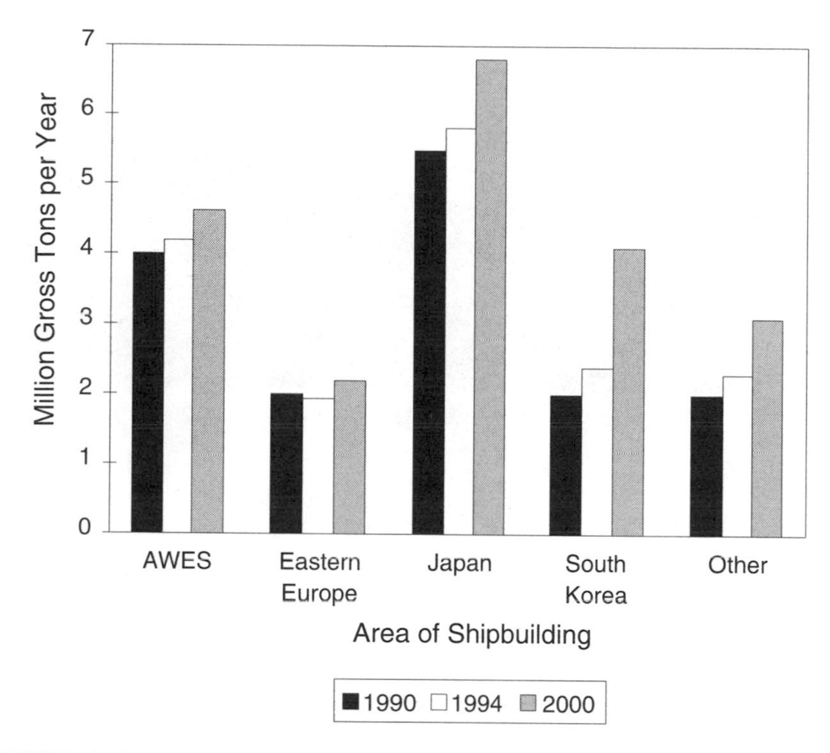

FIGURE 4-15 Increase in shipbuilding capacity by geographic area, 1990–2000. Note: AWES = Association of West European Shipbuilders. Source: de Albornoz, 1996. Reprinted with permission of the Society of Naval Architects and Marine Engineers (SNAME)

increasing since then. Figure 4-15 shows the actual increase in capacity over the period 1990 to 1994 by geographical area and projections to the year 2000.

The expansion of capacity is attributed to several factors, including regular increases in productivity in Western Europe, Japan, and Korea; the creation of several large new yards and additional dry docks in South Korea; and the reentry of Eastern European, former Soviet Union (FSU), and U.S. shipyards into the merchant shipbuilding market. Predictions are that building capacity will exceed demand through 2005 but that excess capacity will fall to a minimum around the year 2000 (see Figure 4-16).

A recent Japanese report (JAMRI, 1995) took a slightly different approach by ascertaining the number of shipbuilding berths and using gross tons rather than compensated gross tons as a measure. Table 4-4 shows the number of building berths or docks for new ships exceeding 40,000 DWT. Most of the new berths or docks are for vessels exceeding 250,000 DWT. Figure 4-17 presents a comparison of shipbuilding supply and demand in gross tons.

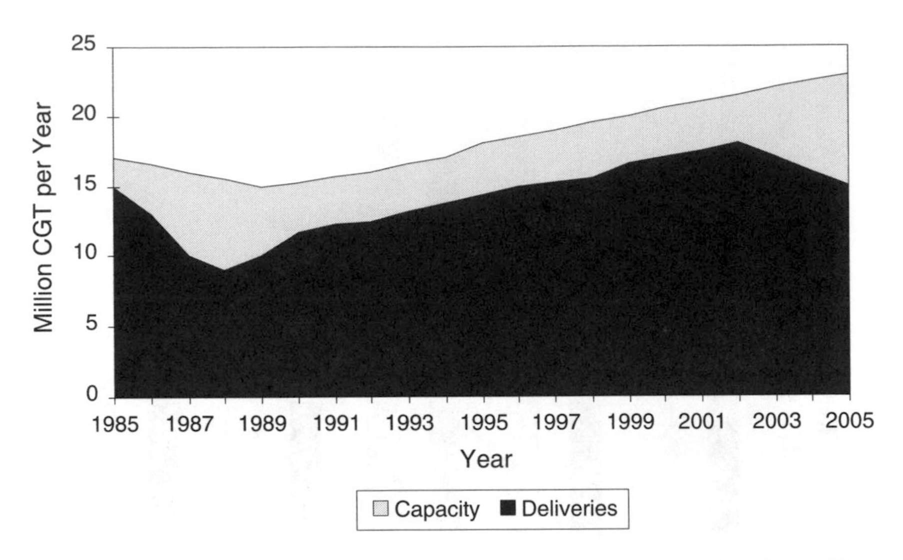

FIGURE 4-16 Comparison of shipbuilding capacity and forecast newbuildings—European estimate. Source: de Albornoz, 1996. Reprinted with permission of SNAME.

Although the recent European and Japanese surveys vary in monetary numbers and dates of peaking, their conclusions are similar: there will be a surplus of newbuilding capacity even when demand peaks at the end of the 1990s.

Availability of Capital

The availability of capital to finance tanker construction will depend on total shipbuilding demand, given that purchasers of all types of vessels compete for capital. Figure 4-18 presents the estimated future capital requirements for new

TABLE 4-4 Number of Shipbuilding Berths or Docks for Vessels Exceeding 40,000 DWT

	Maximum Capacity in 10^3 DWT				
Country	40–80	80–150	150–250	250+	Total Number
Japan	4	9	9	8	30
Korea	2	2 (1)	2	7 (5)	13 (6)
European	17 (–1)	15	6 (2)	10	48 (1)
Others	12	12 (1)	5	4	33 (1)
Total	35 (–1)	38 (2)	22 (2)	29 (5)	124 (8)

Note: Numbers in parentheses represent recent changes in capacity.

Source: JAMRI, 1995.

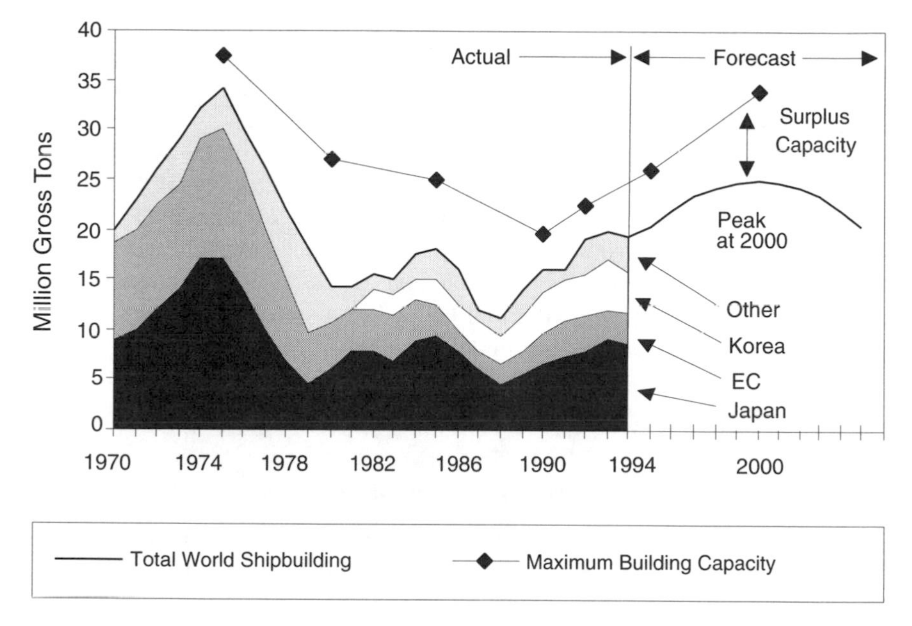

FIGURE 4-17 Comparison of shipbuilding capacity and forecast newbuildings—Japanese estimate. Source: JAMRI, 1995.

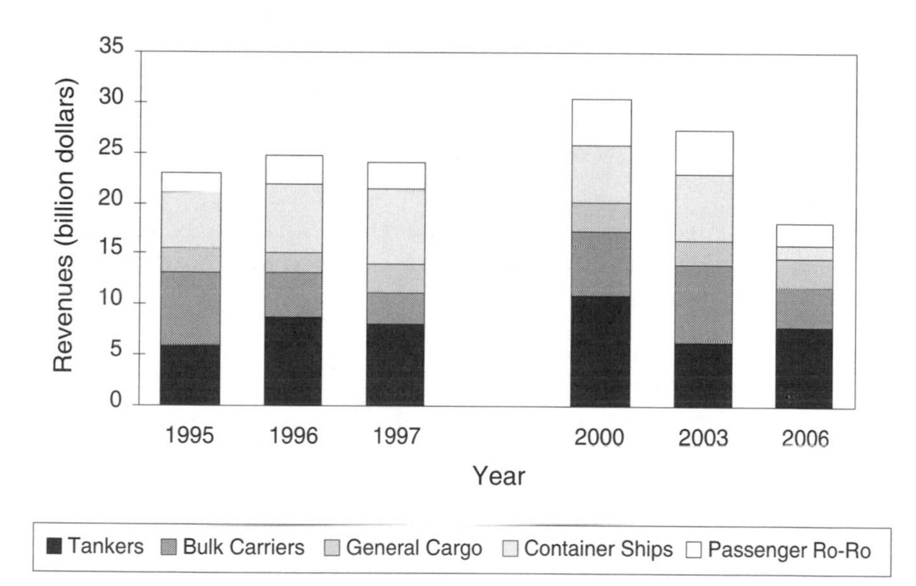

FIGURE 4-18 Estimated shipyard revenues for newbuildings, 1995–2006. Note: Ro-Ro = roll-on and roll-off vessels. Source: Drewry Shipping Consultants, 1996.

ship construction for selected years to 2006. It is estimated that about $35 billion to $47 billion[8] will be needed to finance the new double-hull tankers to be constructed between 1995 and 2000, compared to the $23 billion to $30 billion spent from 1990 to 1995.

Although the demand for financing is expected to peak in the year 2000, the yearly average between 1995 and 2000 will be on the order of $8 billion. Although the demand for capital will be significantly higher than in the first half of the 1990s, the financial community believes that sufficient capital will be available at reasonable interest rates (Newbold and Grubbs, 1995). Reductions in interest rates during the first quarter of 1997 and oversubscription of shipping investment schemes appear to confirm this view.

ECONOMIC COSTS OF OPA 90 AND MARPOL

Incremental Capital Costs

On the basis of a review of industry literature, presentations by various shipbuilders, and questionnaire responses (see Chapter 6), the committee developed a shipbuilding price differential between double-hull and single-hull tankers that is shown in Table 4-5. The increase in cost per DWT for double-hull vessels is estimated at between 9 and 17 percent.

The estimated increase in the cost of building a double-hull fleet of the same size and composition as the tanker fleet in existence on April 1, 1996, is on the order of $12 billion, as shown in Table 4-6. Given a renewal period of 20 years, the annual increase would average about $0.6 billion per year.

Incremental Operating Costs

In comparing the operating costs of double-hull and single-hull tankers of similar size and age, the committee found that maintenance and repair (M&R) and hull and machinery (H&M) insurance premium costs were the only costs that showed marked differences.[9]

Maintenance and Repairs

The only difference between double-hull and single-hull cost in the M&R category is in the cost of maintaining and repairing the protective coatings in the tanks of double-hull vessels. Because such costs are not normally incurred until

[8]Estimates of capital requirements are in 1996 dollars with no inflation imputed.

[9]Insufficient data were available for the committee to quantify potential cost offsets associated with double-hull tanker operations. According to operators surveyed by the committee, double-hull tankers offer reduced times for discharge of cargo and cleaning of cargo tanks (see Chapter 6).

TABLE 4-5 Tanker Newbuilding Prices as of April 1, 1996

		Double Hull		Single Hull		Cost Increase	
Vessel DWT	Tanker Type	($ million)	($/DWT)	($ million)	($/DWT)	($/DWT)	(%)
47,000	Product	33.5	713	30.5	649	64.0	9.9
67,000	Product	40.0	597	36.0	537	60.0	11.1
105,000	Aframax	42.0	400	36.0	343	57.0	16.7
153,000	Suezmax	51.5	337	44.0	288	49.0	17.0
300,000	VLCC	80.5	268	70.0	233	35.0	15.0

later in the life of a tanker, the costs discussed here are total M&R costs for the first 15 years of life. The committee's estimates of M&R costs for double-hull and single-hull VLCC, Suezmax, and Aframax tankers are shown in Table 4-7. The costs for double-hull tankers exceed those for single-hull tankers by 11 to 37 percent, depending on vessel type.

It should be noted here that the committee's estimates of the M&R costs of the two types of vessels depend on the assumption that the added cost associated with double-hull coatings is proportional to the increase in coating area. An earlier NRC study (NRC, 1991) estimated the incremental M&R costs of double-hull tankers to be considerably higher than those estimated here. A more recent study (Whiteside, 1996) asserted that increased vigilance in checking coatings and internal spaces in double-hull tankers would result in the early discovery of problem areas and hence no additional costs for double-hull M&R over the long run.

Insurance Premiums

It is estimated that the costs of marine H&M and war risk insurance per gross ton (GT) for a double-hull VLCC or Aframax tanker are about 6 percent higher

TABLE 4-6 Increased Cost of Building the Double-Hull Fleet

					Cost Increase		
Vessel Type	Size Range (kDWT)	Total Million DWT	Double Hull ($/DWT)	Single Hull ($/DWT)	($/DWT)	(%)	Total ($ billion)
Small	10–60	40.5	713	649	64.0	9.0	2.60
Aframax	60–100	48.6	400	343	57.0	16.7	2.80
Suezmax	100–200	46.9	337	288	49.0	17.0	2.30
VLCC	200+	125.4	268	233	35.0	15.0	4.40
Total		261.4			46.3		12.10

Source: Fleet distribution from Clarkson Research, 1996; cost per DWT from previous table.

TABLE 4-7 Comparison of Maintenance and Repair Costs ($/DWT/year) for Double-Hull and Single-Hull Tankers by Vessel Type

Vessel Type[a]	Double Hull	Single Hull	Cost Increase (% increase)
VLCC	4.89	4.41	0.48 (11)
Suezmax	7.62	5.95	1.67 (28)
Aframax	7.78	5.69	2.09 (37)

[a]M&R differential for small tankers could not be ascertained owing to lack of data.

than for a single-hull tanker of comparable size and age. This difference arises solely from the higher purchase cost of a double-hull vessel.

Discounts in P&I insurance rates are given to all vessels with segregated ballast tanks and vary according to the tanker's age. Token P&I insurance rate discounts given between double-hull tankers in the years 1992 to 1995 were terminated in February 1996. Thus, there are currently no significant differences in P&I insurance costs between double-hull and single-hull tankers. Increases in total insurance costs for various types of double-hull tankers are on the order of 1 percent for VLCCs, 3 percent for Suezmax tankers, and 4 percent for Aframax tankers.

Total Operating Costs

When the increased cost percentages shown above are added to the operating costs for single-hull tankers reported by Drewry (1994), the increase in operating costs attributable to double-hull tankers is estimated at 5 to 13 percent, as shown in Table 4-8. The annual incremental operating cost of a double-hull fleet comparable to the existing fleet is estimated to be on the order of $900 million.

TABLE 4-8 Increase in Operating Costs for Double-Hull Tankers

	Operating Costs ($ thousand/year)		Increases Attributable to Double Hull		DWT of Segment	Cost Increase
	Single Hull	Double Hull	(%)	($/DWT/year)	(10^6 DWT)	($ million)
Product	3,035	3,430	13	9.86	40.5	309
Aframax	3,584	4,050	13	5.18	48.6	252
Suezmax	4,212	4,675	11	3.31	46.9	155
VLCC	5,845	6,137	5	1.04	125.4	131
Total					261.4	937

TABLE 4-9 Incremental Costs of Double-Hull Fleet

	Annualized ($ billion)	20 Year Cycle ($ billion)
Incremental capital cost	0.6	12.1
Incremental operating cost	0.9	18.7
Total incremental cost	1.5	30.8

Table 4-9 summarizes the incremental cost of both constructing and operating a double-hull fleet comparable in size and composition to the existing fleet. The committee estimates the incremental cost to be on the order of $1,500 million per year, or about 10 cents per barrel of oil transported. The 10-cent increment was obtained by dividing the annual incremental cost by ocean trade flows of crude oil and products, which are about 15,700 million barrels (MSI, 1996a). This rough estimate is at the lower end of estimates reviewed by the committee. The NRC study (1991), using the higher prices and costs prevailing at the time, reported an incremental cost of 16 cents per barrel for oil shipments to the United States by double-hull tankers. A more recent study (Brown and Savage, 1996) estimates the incremental cost for U.S. trade to be 14 cents per barrel. That study assumes a loss of cargo capacity resulting from the double-hull structure, which in the opinion of the authors "dictates an increase in fleet size of 2.43 percent." In the committee's judgment, any loss of cargo capacity would be largely offset by effective double-hull designs and would occur only under special circumstances.

Tanker Replacement

The committee conducted an analysis to determine if the phaseout schedules of OPA 90 and MARPOL would result in any changes in the normal ship replacement cycle. The analytical approach used is described in Appendix J. Highlights are presented below.

As a ship gets older, the owner must decide between continued operation of an old, less efficient vessel that may be in need of some repair or replacement by a new, efficient vessel with a high acquisition cost. For the purposes of this analysis, sale of a ship to another owner to operate is not considered;[10] the owner's only options are to operate or to scrap. Therefore, the owner will replace the ship only if (1) the ship is not seaworthy, (2) the tanker cannot be operated economically, or (3) regulations require scrapping. In general, the scrapping age follows a normal (bell-shaped) distribution (Figure 4-19).

[10]In practice, a ship is typically scrapped by its third or fourth owner and not by the original owner.

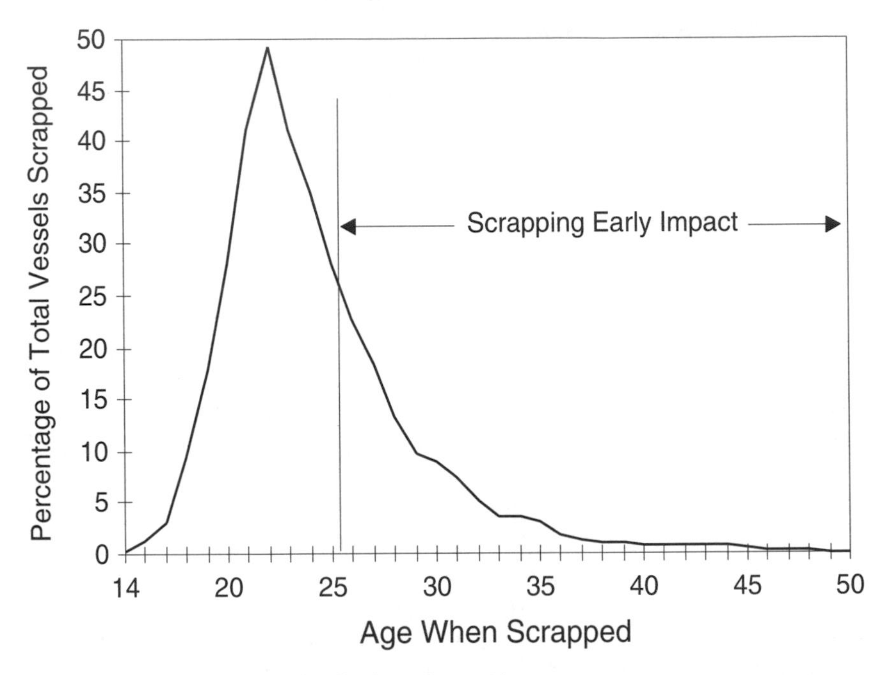

FIGURE 4-19 Generalized distribution of scrapping.

The question is whether Section 4115 will cause some tankers to be scrapped and replaced earlier than the owner would otherwise desire. A vertical line drawn at a particular age on the normal distribution curve illustrates the effect of early scrapping. If ship operation is not allowed (or is too expensive) after this age, the curve is truncated at this point. The tail of the curve to the right of the vertical line indicates the impact of early scrapping.

A number of factors would maximize the impact of early scrapping due to Section 4115:

- Current tankers are in good condition, and the owners wish to operate them longer than the regulatory age restrictions allow.
- There are no non-International Maritime Organization (IMO) trades suitable for such tankers, and they must be scrapped.[11]
- New tankers are much more expensive than they would be without Section 4115.
- New tankers have no operating efficiencies over old tankers.

---

[11] In the committee's analysis, tankers are assumed to meet all the requirements of Section 4115 or 13G. Even if these vessels do not trade to the United States, they are assumed to serve nations that honor 13G.

TABLE 4-10 Break-Even Special Survey Costs ($ million) for Pre-MARPOL Tankers in International Trade

	Capacity Reduction		
Tanker Size (DWT)	0 percent	5 percent	8 percent
40,000	12.7	11.4	10.6
60,000	13.9	12.5	11.7
160,000	21.1	19.2	18.1
280,000	34.2	31.3	29.6

- Section 4115 leads to a shipbuilding boom and the development of new construction facilities.

Factors that would result in little or no impact of early scrapping as a result of Section 4115 include the following:

- Shipowners scrap tankers before they reach the regulatory age restriction. The importance of this factor depends on how many tankers successfully complete their fourth and fifth special surveys.
- Major investments are needed to keep older tankers operating beyond their regulatory age limit, and they are expensive to run.
- Shipowners wishing to extend the life of their tankers beyond the limit defined by Section 4115 find non-IMO trades for them.
- No shipbuilding boom occurs, or if it does occur, enough building capacity exists to keep prices from rising significantly.
- New tankers are much more efficient than old tankers.
- Strict surveys and inspections make it very expensive to operate older tankers.

Table 4-10 shows the break-even cost of special surveys for pre-MARPOL tankers facing their fifth special survey for various losses of capacity due to HBL (using the methods explained in Appendix J). If the special survey is expected to cost more than this break-even amount, the shipowner will scrap the vessel rather than undergo the special survey. Given the high values of these break-even costs, the shipowner is likely to pay for the special survey if the market over the next five years is expected to be favorable.

Most large tankers, a shortage of which might cause a shipbuilding boom, do not survive beyond their fifth special survey (i.e., 25 years).[12] Consequently, it appears that the poor condition of many tankers, combined with poor market

[12]According to Clarkson Research, from 1990 to 1994, upward of 90 percent of the VLCCs that reached the fifth special survey were scrapped, as were a third of those reaching the fourth special survey (Clarkson, 1996a).

rates, has had a greater impact on tanker replacement than Section 4115. Section 4115 will have little or no impact on the replacement of large tankers that utilize HBL and are suitable for unloading within U.S. lightering zones or at the deepwater port. If the tanker is in reasonable condition it will pass the fifth special survey, and if future freight rates appear favorable, it will be operated for another five years. The higher the freight rate, the greater is the inducement to extend the tanker's life with HBL.

The situation is different for tankers of less than 120,000 DWT. At least some of these could have passed their fifth special survey and operated economically for another five years with HBL. However, because Section 4115 does not include a life extension for vessels using HBL—in contrast to 13G—these vessels will be effectively banned from U.S. trade[13] and will be either forced into early scrapping or restricted to non-U.S trade.

FINDINGS

Finding 1. The additional construction and operating costs of replacing single-hull tankers by double-hull tankers are expected to total about $30 billion worldwide over a 20-year period. The construction costs of double hulls are expected to run between 9 and 17 percent higher than the construction costs of new single hulls, whereas operating costs are expected to be 5 to 13 percent higher.

Finding 2. Regulations mandating a transition to double-hull vessels are unlikely to result in the withdrawal from service of single-hull vessels in international trade before the end of their economic life, for the following reasons in particular:

- Pre-MARPOL tankers can extend their lives in international trade by up to five years—although not in trade to the United States—without capital investment by using HBL.
- By using the Louisiana Offshore Oil Port (LOOP) or a designated lightering area, single-hull tankers in international trade can continue to operate to the United States until 2015. For economic reasons, only tankers in excess of 150,000 DWT are likely to use this option.
- On the basis of historical trends, many tankers are likely to be scrapped before their statutory retirement dates. However, the historical scrapping pattern reflects an extended period of oversupply and depressed markets; these conditions are likely to change between 2000 and 2005.

Finding 3. The world's shipbuilding capacity has expanded in the past several years and will be sufficient to meet the demand for construction of new double-hull tankers. However, a possible short-term bunching of orders would increase

[13]Smaller tankers are assumed not to use the Louisiana Offshore Oil Port (LOOP) or lightering zones.

the construction prices of all new ships and not just tankers. If such a situation were to arise, the economic impacts of Section 4115 and MARPOL would be greater than those estimated in this chapter.

Finding 4. Although sufficient capital is available to finance the replacement of single-hull by double-hull tankers, present market conditions have not stimulated such newbuilding because (1) prevailing freight rates are inadequate to provide the resources needed for new double-hull vessel construction; (2) there is no significant rate premium for double-hull vessels; and (3) a preponderance of spot (short-term) fixtures and a corresponding dearth of time (long-term) charters has made many shipowners less willing to invest in new tonnage. However, tanker ordering is often driven by market sentiment rather than by purely economic factors, and investment in new tankers is intrinsically unpredictable in its timing.

REFERENCES

Brown, R. Scott, and Ian Savage. 1996. The economics of double-hulled tankers. Maritime Policy and Management 23(2): 171.

Clarkson Research, Ltd. 1995. Tanker Register. London: Clarkson Research.

Clarkson Research, Ltd. 1996a. Shipping Intelligence Weekly July 5.

de Albornoz, Carlos. 1996. The international shipbuilding market. Proceedings of the Society of Naval Architects and Marine Engineers (SNAME) 1996 Ship Production Symposium, San Diego, Calif., February 14, pp. 5–13. New York: SNAME.

Drewry Shipping Consultants. 1994. The International Oil Tanker Market: Supply, Demand and Profitability to 2000. London: Drewry Shipping Consultants.

Drewry Shipping Consultants. 1996. The Shipbuilding Market Analysis and Forecast of World Shipbuilding Demand, 1995–2010. London: Drewry Shipping Consultants.

Energy Information Administration (EIA). 1996. Annual Energy Outlook 1996. Washington, D.C. : U.S. Department of Energy.

Japanese Maritime Research Institute (JAMRI). 1995. Recent Trends of China's Shipping and Shipbuilding. Tokyo: Japan Maritime Research Institute.

Jones, Samuel. 1995. Presentation to the Committee on Oil Pollution Act of 1990 (Section 4115) Implementation Review. Irvine, California, June 12.

Loucas, John. 1995. Presentation to the Committee on Oil Pollution Act of 1990 (Section 4115) Implementation Review. Irvine, California, June 12.

Marine Strategies International (MSI). 1996a. Oil & Tanker Market Update: New Lease on Life. London: Marine Strategies International.

MSI. 1996b. Freight Forecaster. July 1996. London: Marine Strategies International.

National Research Council (NRC). 1991. Tanker Spills: Prevention by Design. Marine Board, Washington, D.C.: National Academy Press.

Navigistics Consulting. 1996. Tanker Supply and Demand Analysis, Task 1 and Task 2. Study prepared for the Committee on Oil Pollution Act of 1990 (Section 4115) Implementation Review. Washington, D.C., June.

Newbold, John L., and James L. Grubbs. 1995. Presentation to the Committee on Oil Pollution Act of 1990 (Section 4115) Implementation Review. Washington, D.C., October 9.

PIRA Energy Group. 1995. Forecast of Oil Demand. Prepared for the Committee on Oil Pollution Act of 1990 (Section 4115) Implementation Review. New York: PIRA Energy Group.

PIRA Energy Group. 1996. Energy Briefing. Presentation to Mr. Ran Hettena. New York. June 18.

Shawyer, Eric F. 1995. Presentation to the Committee on Oil Pollution Act of 1990 (Section 4115) Implementation Review. Washington, D.C., October 9.

Stopford, Martin. 1996. The Oil Tanker Industry, The Last 25 Years in Review. London: Clarkson Research.

Tamvakis, Michael N. 1995. An investigation into the existence of a two-tier spot freight market for crude oil carriers. Maritime Policies and Management 22(1): 81–90.

Whiteside, Richard. 1996. Presentation to the Committee on Oil Pollution Act of 1990 (Section 4115) Implementation Review. Irvine, California, February 1.

BIBLIOGRAPHY

Cambridge Energy Research Associates. 1995. World Oil Trends. Cambridge, Mass.: Cambridge Energy Research Associates.

———. 1995. Refined Products Watch. Cambridge, Mass.: Cambridge Energy Research Associates.

———. 1996. World Oil Trends. Cambridge, Mass.: Cambridge Energy Research Associates.

———. 1996. Refined Products Watch. Cambridge, Mass.: Cambridge Energy Research Associates.

Energy Security Analysis, Inc. 1995. Pacific Basin Stockwatch Market View. Washington, D.C.: Energy Security Analysis, Inc.

———. 1996. Pacific Basin Stockwatch Market View. Washington, D.C.: Energy Security Analysis, Inc.

Japanese Maritime Research Institute. 1992. Trends of the World Shipping and Shipbuilding in 1991 and Prospects for the Same in the Near Future. JAMRI report No. 43. Tokyo: Japan Maritime Research Institute.

———. 1994. Recent Changes in the South Korean Shipbuilding Industry and Its future Prospects. JAMRI Report No. 49. Tokyo: Japan Maritime Research Institute.

Morgan, J.P. 1995. Overview of the financing outlook for the tanker industry. Prepared under the direction of James Hamilton (J.P. Morgan, Inc.) and distributed to the Committee on Oil Pollution Act of 1990 (Section 4115) Implementation Review at third committee meeting, Washington, D.C., October 9.

Stopford, Martin. 1990. The supply, demand, and freight rates in the bulk shipping market. Presented at Shipping 90 Conference, Stamford, Connecticut, March 19.

5

Domestic (Jones Act) Tanker Trade

This chapter examines the economic viability of the domestic (Jones Act) petroleum transportation industry[1] in light of the double-hull requirements of Section 4115 of the Oil Pollution Act of 1990 (OPA 90). Under the terms of what is commonly called the Jones Act (the formal name being the Merchant Marine Act of 1920),[2] shipping between any two points in the United States is restricted to U.S.-registered vessels owned by U.S. citizens and crewed by U.S. seafarers, built in the United States without construction differential subsidies (CDS) and operated without operating differential subsidies (ODS).[3,4] Accordingly, foreign tankers are precluded from operating in the U.S. domestic trade.

The domestic tanker industry carries approximately 4 million barrels per day of oil (MBD), equivalent to about half the amount imported into the United States. Domestic trade consists of the Alaskan crude oil trade and the coastal products trade. The Alaskan crude oil trade utilizes large tankers (50,000 to 265,000 DWT)

---

[1]The committee's assessment is limited to the coastal (oceangoing) U.S. maritime transportation industry. The extensive inland and intracoastal petroleum transportation network using tank barges is, therefore, beyond the scope of the present analysis.

[2]1, 41 Stat.988, Chapter 250 of Statutes at large.

[3]Crude oil moving from Alaska to the U.S. Virgin Islands has been exempted from Jones Act restrictions and can be carried in foreign tankers. About 6 percent of the Alaska-trade, or 86,000 barrels per day, was carried to the Virgin Islands in 1994. U.S.-flag tankers operating in foreign trade are considered part of the international trade and are not included in the analysis and discussion in this chapter.

[4]Construction differential and operating differential subsidies are U.S. government programs that offered subsidies to shipowners and shipbuilders to allow them to compete in world trade, effectively reducing the operating and capital cost of U.S.-built ships to world levels. These programs are not currently available for new tankers.

hauling Alaskan crude oil from the terminus of the Trans-Alaska Pipeline System (TAPS) to the U.S. west coast, Panama (for transshipment to Gulf and east coast refineries), and Hawaii. As of June 1996, the export of crude oil from the Alaskan North Slope (ANS) to foreign countries has been permitted, provided the oil is carried on U.S. flag vessels.

The coastal products trade primarily involves the movement of petroleum products from Gulf coast refineries to east coast distribution terminals. The coastal trade also includes the movement of methyl tertiary-butyl ether and petroleum blend and feed stocks from the Gulf to the west coast, the transportation of petroleum products along the west coast, and the shipment of petroleum products from refineries in the U.S. Virgin Islands and Puerto Rico. The vessels in these trades are tankships of 18,000 to 50,000 DWT and coastal barges and integrated tug-barges (ITBs) of less than 10,000 DWT up to 50,000 DWT. Small crude tankers (50,000 to 70,000 DWT) also operate in the coastal products trade, depending on market conditions. The committee's analysis assumes that these vessels are available for ANS trade but not for the coastal products trade.

Jones Act tank vessel trade is highly competitive.[5] The market is characterized by a competitive structure and behavior with low barriers to entry. In addition to competition with other vessels, tank vessels in the domestic trade compete with the highly developed land-based pipeline transportation system in the United States and with foreign tankers carrying refined petroleum products to U.S. ports.

The U.S. Maritime Administration (MARAD), part of the U.S. Department of Transportation (DOT), is charged with promoting the U.S. merchant marine and shipbuilding industry. This is accomplished through a variety of programs, such as Title XI financing guarantees.[6] MARAD is responsible for overseeing the Jones Act trades and ensuring that the supply of vessels is adequate. MARAD has the authority to grant waivers allowing noneligible vessels to operate on domestic routes. These waivers are issued to prevent any interruption of service and to moderate increases in freight rates when a shortage of vessels occurs. For example, MARAD granted waivers for limited periods in the late 1970s and early 1980s that allowed tankers built with construction differential subsidies to enter the Alaskan crude oil trade.

Jones Act tank vessels are subject to the retirement provisions of Section 4115 of OPA 90, but they are not subject to the retirement provisions of the MARPOL Regulation I/13G (MARPOL 13G), which the United States has not ratified.

---

[5]Low or nonexistent profitability and excess capacity are not inconsistent with a competitive market. A highly competitive market can lead to "excessive competition" in the case of chronic maladjustments arising from a combination of circumstances, such as a highly competitive, unconcentrated industry; very easy entry; very slow exit in the case of overcapacity because the productive plant is long-lived and not convertible to alternative uses; and declines in demand for the industry output over time (Bain, 1959).

[6]Title XI is a financing guarantee program in which the U.S. government guarantees payment on loans used to construct vessels in U.S. shipyards.

TABLE 5-1 Jones Act Tank Vessel Fleet by Hull Type

Tank Vessel Type	All Hull Types (number)	All Hull Types (DWT)	Double Hull (number)	Double Hull (DWT)
Tankers > 50,000 DWT	44	4.8	3	0.4
Tankers < 50,000 DWT	72	2.7	20	0.7
Coastal barges	93	1.7	17	0.4
ITBs	13	0.5	3	0.1
Total	222	9.7	43	1.6

TANK VESSEL SUPPLY

As of August 1995, the Jones Act fleet consisted of 222 tankers, tank barges, and ITBs with a total capacity of 9.7 million DWT.[7] Single-hull vessels that are subject to retirement by 2015 under OPA 90 account for 85 percent of the total deadweight capacity. The composition of the fleet by hull type is shown in Table 5-1.

Figure 5-1 shows the construction dates of vessels in the fleet, which is considerably older on average than the world fleet (see Figure 4-2). The peak years of construction reflect the opening of TAPS.

Alaskan Crude Oil Trade

The fleet of Jones Act crude oil tankers exceeding 50,000 DWT consists of 44 tankers with a total capacity of approximately 4.8 million DWT. Only three of these tankers have double hulls;[8] the other 41 will be phased out of service under OPA 90. The available supply shown in Figure 5-2 is based on the OPA 90 retirement schedule (i.e., no retirements for economic or other reasons are included). The supply figures do not include four 165,000 DWT tankers controlled by British Petroleum (BP) and operated under the Jones Act by Keystone Shipping Company and Interocean Management that were withdrawn from service in 1995. It is uncertain whether BP will fit double-hull forebodies on the existing ships or build new double-hull tankers. Two rebuilt 125,000 DWT double-hull vessels controlled by BP are included in the tanker supply beginning

[7]Data on this trade were developed from MARAD, Clarkson's tanker database, and vessel operators involved in the trade (primarily British Petroleum, Maritime Overseas Corporation, and Keystone Shipping). Unless otherwise noted, figures and tables in this chapter were developed by the committee using these sources.

[8]Sun Shipbuilding's Environmental Class of 120,000 DWT tankers built in the 1970s—the *Tonsina, Kenai*, and *Prince William Sound*—are the only double-hull tankers of more than 50,000 DWT in the Jones Act trade.

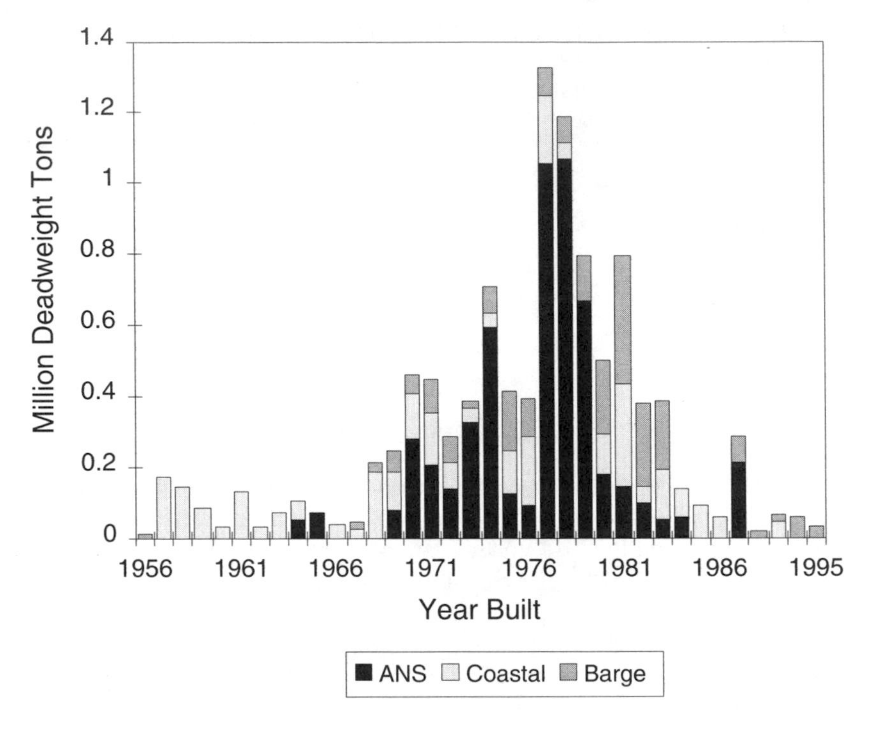

FIGURE 5-1 Age profile of Jones Act fleet.

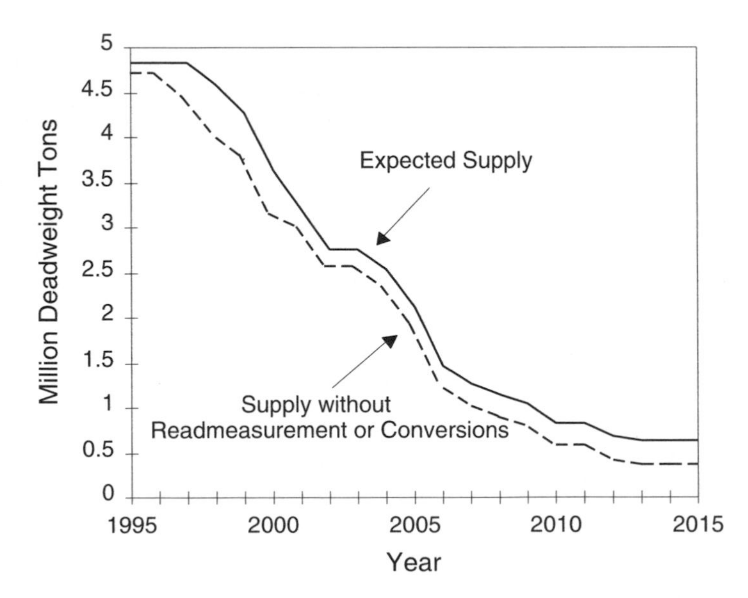

FIGURE 5-2 Jones Act tank vessel supply (vessels of more than 50,000 DWT).

in 2000. It is assumed that no other tankers will be added to the supply of vessels exceeding 50,000 DWT. The supply also includes the potential impact of readmeasurement. This impact is assumed to be modest, based on identification of likely candidates for readmeasurement by the U.S. Coast Guard (USCG) and MARAD.[9] The lower curve in the figure shows the potential supply without conversions or readmeasurement.

Coastal Products Trade

Supply of Jones Act Tankers of Less Than 50,000 DWT

According to MARAD, there are 72 tankers totaling 2.7 million DWT in the coastal products trade,[10] of which 20 tankers (with a total capacity of 700,000 DWT) have double hulls. Four American Heavy Lift vessels that are presently being converted to double hulls at Avondale Shipyard and five Double Eagle double-hull tankers under construction for Hvide Van Ommeren Tankers L.L.C. at Newport News Shipbuilding are included in the supply.[11] Figure 5-3 shows the impact of retirements on the tanker fleet of less than 50,000 DWT through 2015. Also shown is the potential impact of readmeasurement and the entry of tankers originally constructed with CDS that will likely be able to enter the Jones Act trade after reaching 20 years of age.

Supply of Tank Barges and ITBs

According to MARAD, 93 barges exceeding 5,000 gross tons (GT) will be affected by Section 4115 of OPA 90. Seventeen are double hulls that will not be retired pursuant to OPA 90. The remaining 76 are single hulls that will be phased out by 2010. There are also 13 ITBs, of which three have double hulls. The supply of tank barges and ITBs through 2015 is shown in Figure 5-4.

---

[9]OPA 90's retirement schedule is based on the gross tonnage (GT) and build date of a vessel. It is possible to readmeasure a vessel to reduce its GT below the GT categories in OPA 90 to gain between one and five years of additional service. As of June 1997, USCG records indicate that few vessels have been readmeasured to reduce their GT (personal communication from Jaideep Sirkar, USCG, to Ran Hettena, member, Committee on OPA 90 [Section 4115] Implementation Review, June 1997).

[10]MARAD includes four tankers that were built with CDS but, having reached 20 years of age, are now considered eligible to operate in Jones Act trade.

[11]The first of four 45,000 DWT product carriers to be built in a U.S. shipyard for foreign-flag operation has recently been resold to a major U.S. oil company for domestic service. Thus, the U.S.-flag double-hull tanker capacity is higher than that projected in the committee's analysis. It is unclear whether any of the three remaining tankers will also be sold for Jones Act operation.

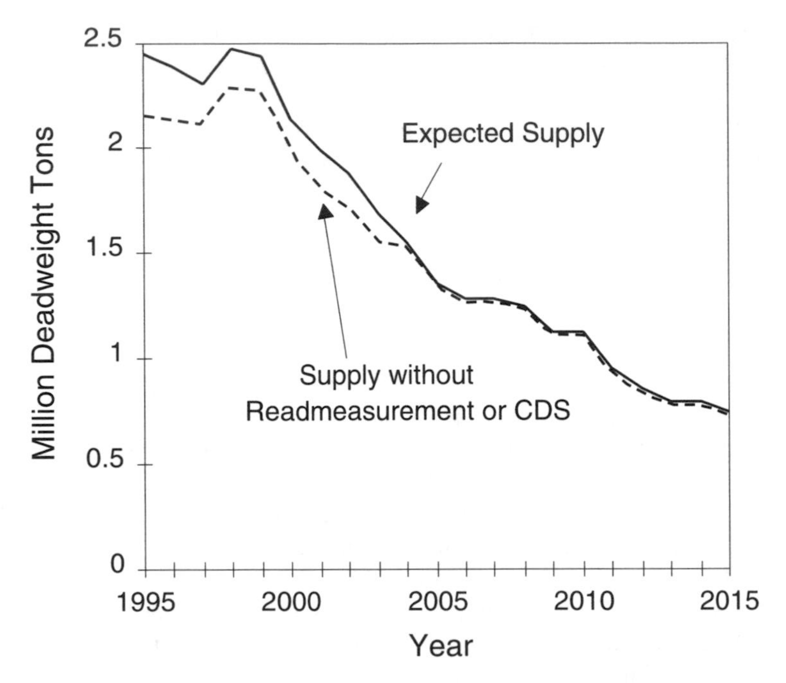

FIGURE 5-3 Jones Act tank vessel supply (vessels of less than 50,000 DWT).

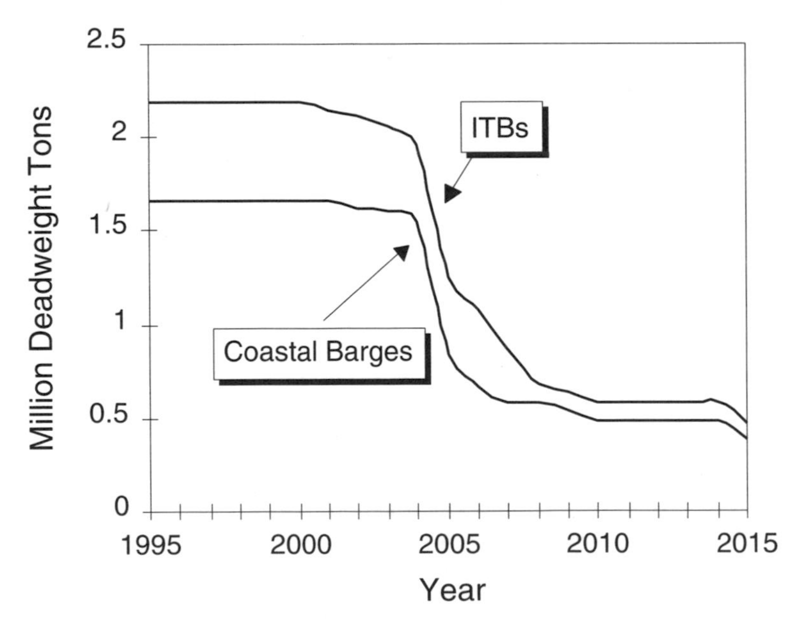

FIGURE 5-4 Jones Act tank vessel supply (tank barges and ITBs).

SUPPLY-DEMAND BALANCE

ANS Crude Oil Tanker Trade

The key factors in projecting the demand and supply of vessels used in the ANS crude trade are projected ANS production, tanker fleet capacity utilization, and exports of ANS crude to the Far East.

Estimates of ANS production over the next 20 years vary greatly, as shown in Figure 5-5. The Energy Information Administration (EIA) Reference case is the lowest estimate, PIRA's, the highest. The Alaska Department of Revenue (ADR) projection falls between these two estimates, except after about 2010. Both the EIA Reference case and the EIA High Oil Price case show increasing production after the 2010–2012 time period, but ADR shows a continuing decline (State of Alaska, 1995). This latter ANS production forecast assumes that no new fields are brought on-line, that the Alaskan National Wildlife Refuge is not successfully developed, and that existing high marginal cost fields, not on-line, remain off-line.

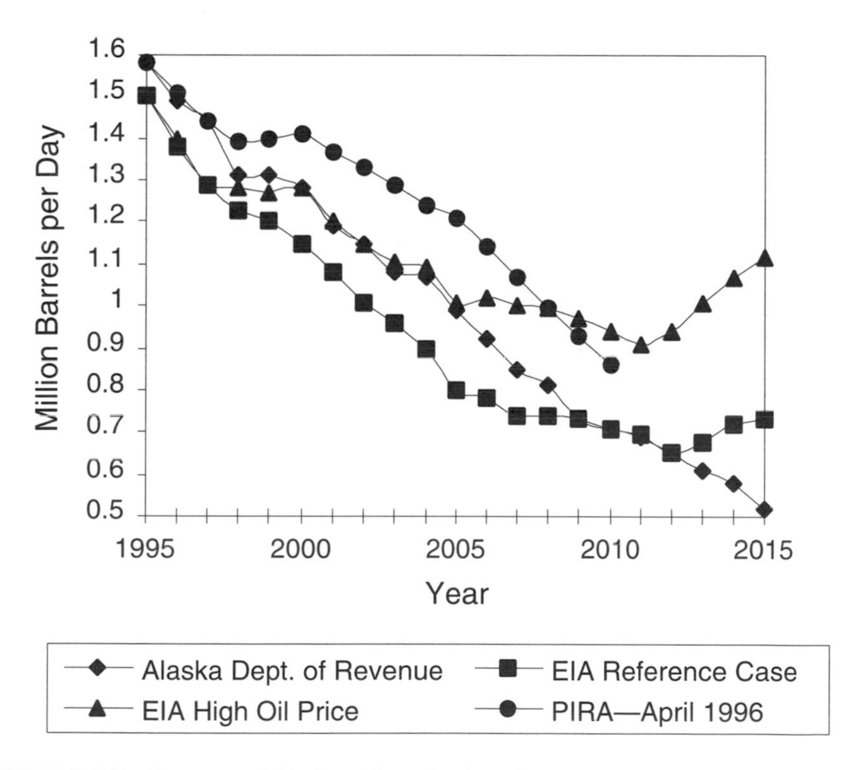

FIGURE 5-5 Forecasts of Alaskan oil production. Sources: EIA, 1996; State of Alaska, 1995; PIRA, 1996.

Each of the three major producers in Alaska—Exxon, BP, and ARCO—uses its own proprietary fleet,[12] with additional tonnage chartered when necessary. Under these circumstances, the adequacy of tonnage of each company depends to some degree on how efficiently it uses its own fleet. Since the passage of OPA 90, some producers have been reluctant to use their own ships to carry oil for others because of liability concerns. This may result in lower productivity when one producer has excess tonnage available while another is short. Nonetheless, utilization rates can be expected to rise as the supply of available ships tightens, thus delaying any tanker shortage.

Exports of ANS crude have increased the complexity of efforts to project U.S. tanker demand. These exports may or may not increase the demand for Jones Act vessels, since the law requires only that vessels carrying ANS oil be American owned and flagged. There is no requirement that they be built in U.S. shipyards or crewed by U.S. seafarers. However, the biggest ANS producer and probable principal exporter—BP Oil, Inc.—has committed to using Jones Act vessels. In a letter to Senator Dianne Feinstein (141 Cong. Rec. S6665, May 15, 1995), BP Oil, Inc. stated that "BP will commit now and in the future to use only U.S.-built, U.S.-flag, U.S.-crewed ships for [Alaskan oil] exports. We will supplement or replace ships required to transport Alaskan crude oil with U.S.-built ships as existing ships are phased out under the provisions of the Oil Pollution Act of 1990."

Exports of ANS crude are just starting, and there are widely divergent projections of their volume. The minimum amount likely to be exported is any production exceeding the West Coast requirement of about 1.3 MBD. Estimates range from a low of 20,000 barrels per day (b/d) to more than 300,000 b/d. Government forecasts are on the order of 150,000 b/d. The committee's analysis relies on the PIRA figures shown in Table 5-2.

TABLE 5-2 Projected ANS Crude Oil Exports

Year	Exports to Far East (b/d)
1996	15,000
1997	200,000
1998	190,000
1999	190,000
2000	180,000
2005	70,000
2010	No exports

Note: All ANS crude oil exports are assumed to go to the Far East.

Source: PIRA, 1995.

[12]These fleets include ARCO's and BP's long-term chartered or leverage leased vessels.

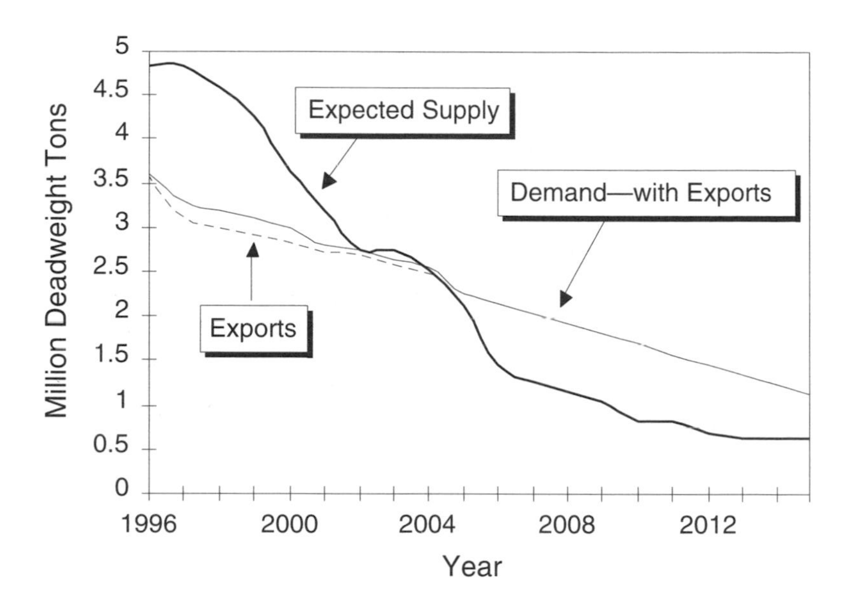

FIGURE 5-6 ANS crude oil trade supply and demand. Sources: State of Alaska, 1995; Navigistics, 1996.

The supply-demand balance of vessels for the ANS crude trade is based on relatively conservative forecasts of ANS production and exports by the ADR (Figure 5-6). Supply consists of all Jones Act tankers exceeding 50,000 DWT with the assumption of high efficiency in utilization. Under these conditions, demand equals supply in 2002 and continues to do so until 2006, when significant new vessel construction of about 850,000 DWT would be needed.

In Figure 5-7, the PIRA estimate is overlaid on the ADR projection adopted by the committee and shows a significant vessel shortfall—between 300,000 and 500,000 DWT—starting in about 2002.

The overriding difficulty in offsetting the projected decline of the fleet through new construction is the industry's expectation of a decline in ANS crude production. Such a decline is not certain. New exploration successes, lower production costs, improved recovery techniques, or the opening of the Arctic National Wildlife Refuge to oil production might prevent it. These developments aside, the possibility that demand may disappear after only half the economic life of new double-hull vessels is exploited weighs heavily against the incentive to replace. Figure 5-8 illustrates the problem. Beginning in 2004, newbuildings will be needed to replace tankers phased out through 2010. By 2011, the expected decline in production will result in surplus capacity among vessels constructed during the previous six years. Approximately 40 percent of the new vessels will not be in use by 2015, less than 10 years after they were built. Given this forecast,

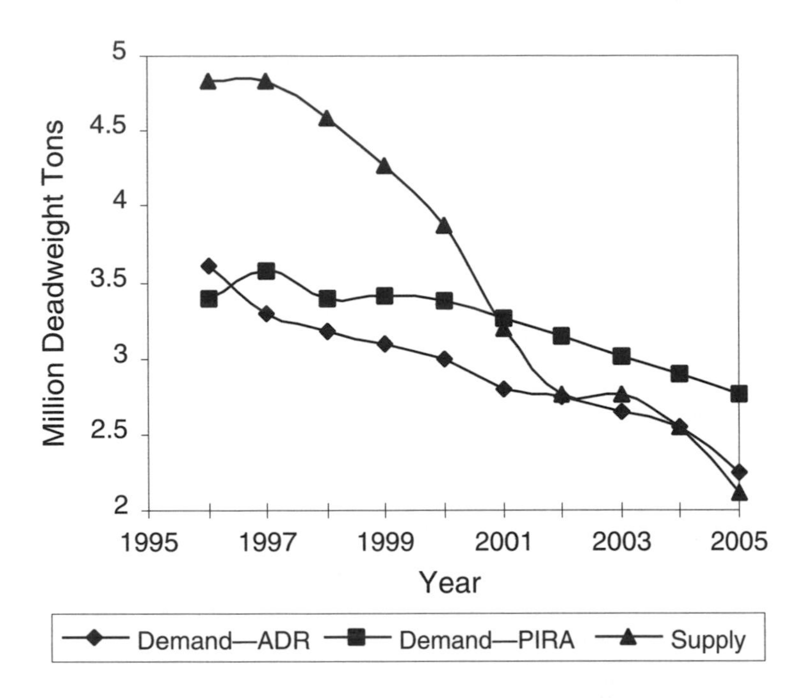

FIGURE 5-7 ANS crude oil supply and demand with alternative demand forecast, 1996–2005. Sources: Navigistics, 1996; PIRA, 1996; State of Alaska, 1995.

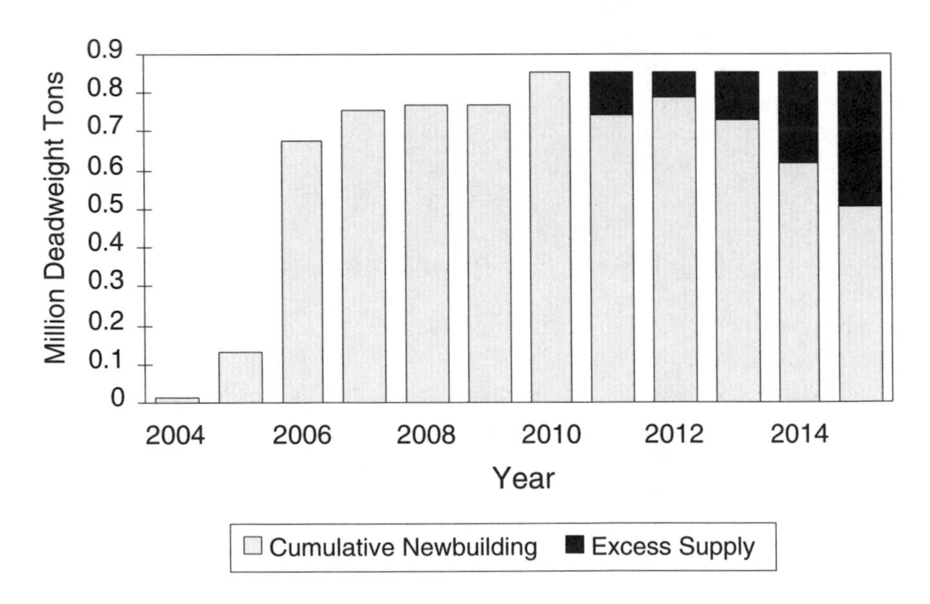

FIGURE 5-8 ANS newbuilding conundrum.

TABLE 5-3 Product Supply Methods to the Eastern United States (MBD), 1993

Area	Pipeline	Jones Act Vessels	Foreign Imports
Florida	—	0.68	0.12
South Atlantic	1.15	0.15	0.08
Mid-Atlantic	0.98	0.11	0.36
Northeast	—	0.63	0.25

Source: Wilson, Gillette, & Co., 1994.

the present slow pace of new orders is not likely to accelerate markedly in the near future.

Under the circumstances, rebuilding may offer an attractive alternative to new construction. Equipping existing single-hull forebodies with double hulls may turn out to be more economically feasible. In either case, U.S. shipbuilding capacity is adequate to meet the need, as shown later in this chapter.

Coastal Products Trade

Domestic tank vessels used in the east coast products trade face different and more complex competitive problems than those affecting international transportation.[13] These vessels compete not only among themselves, but also against oil pipelines and (indirectly) against foreign ships carrying imports (see Table 5-3).

The pipeline system is elaborately reticulated and flexible, and its cost and rate structure make it cheaper than tank vessels. In the South and in mid-Atlantic markets, it is the carrier of choice for product shipments from the Gulf to the east coast, accounting for almost 90 percent of trade. Because pipeline capacity is not fully utilized for at least six months each year, the pipeline acts as an effective ceiling on vessel tonnage during that period.

Unusually severe winter weather or refinery shutdowns, coupled with shortages induced by just-in-time inventory policies, occasionally expand opportunities for coastwise tankers, which have a time advantage over imports when delivery is urgent. Pipelines, however, use their spare capacity during the low-demand season to build up distribution-point inventories for use in the high-demand season. Furthermore, expansion of pipeline capacity over the long run is possible.

[13]An assessment of Jones Act tanker demand in this trade was made in a study by Wilson, Gillette & Co. (1994) for MARAD in anticipation of the implementation of reformulated gasoline regulation. The study forecasts a growth in the coastwise tank vessel trade from 3.5 million DWT in 1994 to 4.15 million DWT in 2005. In January 1996, MARAD's Office of Statistical and Economic Analysis completed a limited assessment that concluded: "The market can accommodate nine additional product tankers and rates will rise no later than the turn of the century" (MARAD, 1996).

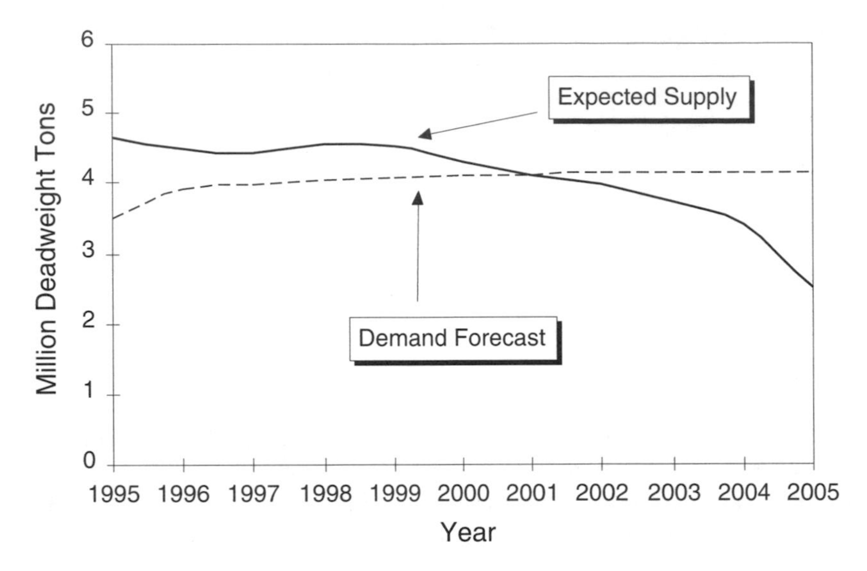

FIGURE 5-9 Jones Act tank vessel coastal product supply and demand. Sources: Wilson, Gillette & Co., 1994; Navigistics, 1996.

Although the Jones Act insulates the coastwise trade from foreign ship competition between U.S. ports, it does not preclude direct imports of competitive products in foreign vessels to U.S. ports. In fact, the volume of product imports from foreign refineries can be directly induced by the level of U.S. rates.[14] The comparatively low capital base of the domestic fleet due to its age strengthens its ability to hold its present market share. Additionally, the market has witnessed a chronic excess supply of vessels that has depressed the freight rates of all tankers and coastal barges for more than a decade. Moreover, pronounced seasonal variations in shipments cause freight rates to fall still lower during much of the year (see Gassman, 1996; MARAD, 1996).

The flow of products from the Gulf to the East Coast has declined as production has moved from Gulf Coast refineries to the Midwest. East Coast demand for products is being met by foreign imports. As a result, there is reduced demand for Jones Act vessels to move products from the Gulf to the East Coast.

The demand for waterborne petroleum products was fairly level in the early 1990s and is forecast to grow moderately, as shown in Figure 5-9. The future course of Jones Act tanker demand, however, is in question.

Because the Gulf and East Coast tanker and coastal barge market is limited

[14]MARAD (1996) observes that "at time charter rates greater than $30,000 per day for Jones Act product tankers, imports on cheaper foreign flag tankers would increase sharply at the expense of domestic shipment."

to what the pipelines cannot carry and is also subject to competition from foreign imports, the extent to which single-hull vessels will be replaced is hard to predict. As older vessels are phased out, they will be replaced only to the extent that double-hull tankers and barges obtain freight rates commensurate with their higher construction costs. If these rates significantly exceed prevailing and anticipated future market rates, however, the pipelines and imported oil are likely to gain market share at the expense of domestic tank vessels. Under such conditions, rates are likely to be perceived as too low to support the higher cost of new or converted double-hull tankers or barges. Some new construction will be needed to carry oil products to regions that cannot be reached by pipeline or served by foreign imports, but a significant part of the single-hull coastal fleet may not be replaced as it is phased out. A major reduction in the capacity of the U.S.-flag tanker fleet and increased dependence on foreign imports could increase concerns over national security issues.

SHIPYARD CAPACITY AND AVAILABILITY OF CAPITAL

In the United States there are currently 21 shipyards with 123 building slots capable of building or reconstructing double-hull Jones Act tankers (MARAD, 1995). The capacity of these shipyards is shown in Table 5-4.

By assuming that a double-hull tanker can be constructed within two years of keel laying, the capacity of U.S. shipyards is adequate to cover the projected shortfall in Jones Act tanker capacity in the ANS trade, provided construction begins in 1998. Converting an existing tanker into a double hull will likely take less time and would overcome the projected shortfall if construction started as late as 1999. If demand exceeds current projections, newbuilding or reconstruction would have to start earlier. Since U.S. shipyard capacity is substantial, the entire existing domestic fleet of single-hull tank vessels of less than 50,000 DWT

TABLE 5-4 Number of U.S. Industry Vessel Building Slots[a]

	Number of Building Slots by Vessel Size			
Coast	25,000 DWT	38,000 DWT	89,000 DWT	120,000 DWT
East	16	12	5	2
Gulf	32	26	6	5
West	7	6	4	2
Total	55	44	15	9
Potential DWT	1,375,000	1,672,000	1,335,000	1,080,000

[a]Shipbuilding yards commonly have more than one building berth or building slot.

Source: MARAD, 1995

could probably be replaced within two years, as long as shipyard capacity was not constrained by demand for naval or other types of commercial vessels.

Members of the financial community (Morgan, 1995; Newbold and Grubbs, 1995) stated in presentations to the committee that there are no restrictions on the availability of capital for economically viable construction or reconstruction of double-hull tank vessels for the Jones Act trade. In addition, MARAD is currently authorized to provide financing guarantees (through the Title XI program) for the construction or reconstruction of double-hull tankers in U.S. shipyards.

ECONOMIC IMPACT OF SECTION 4115 ON DOMESTIC SHIPPING

Capital and Operating Costs

The differences in the capital and operating costs of double-hull and single-hull vessels in Jones Act trade are similar to those found in the international fleet. Increases in operating costs attributable to double-hull replacement are estimated to be between 5 and 13 percent. Increases in capital cost for double-hull compared to single-hull vessels are approximately 10 to 17 percent. Estimates of the differential cost factor for double-hull tankers built in U.S. compared to foreign yards range from about 1.5 for vessels of 40,000 DWT to as much as 2.25 for vessels of more than 140,000 DWT.[15] Given market uncertainties, the committee did not attempt to estimate the total incremental cost of constructing and operating a double-hull Jones Act fleet comparable in size and composition to the existing fleet.

Required Freight Rates

MARAD (1996) has estimated that a new 45,000 DWT double-hull tanker in the coastal products trade must earn more than $30,000 per day in full-time employment to break even. Prevailing rates in the trade are about $17,000 to $22,000 per day.

Information on specific rates in the Alaskan trade was not available to the committee because of the dominance of proprietary and time-chartered vessels and the consequent absence of a spot market and associated freight rates. Therefore, a comparison could not be made of prevailing rates and rates that would be required to pay for new tankers.

Calculations of Ship Replacement Costs

The formulas used to analyze ship replacement cost in the international trade also apply to domestic trade. There are some significant differences, however.

[15]Personal communication from Keith Michel, Herbert Engineering Corporation, and David St. Amand, Navigistics Consulting, to Marine Board staff, Washington, D.C., June 1997.

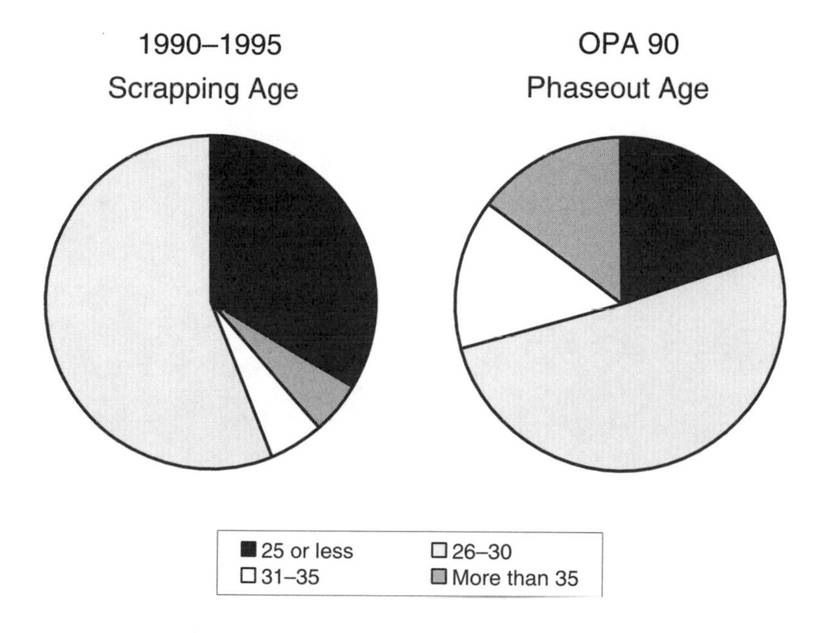

FIGURE 5-10 Comparison of historical scrapping pattern and OPA phaseout age for Jones Act tanker fleet. Source: Clarkson Research, 1990–1995.

Jones Act tankers generally cost more to build than tankers for international trade. In addition, freight rates are low because of excess capacity. As a consequence, the owners of vessels in the domestic trade typically operate their ships for 20 to 35 years or more, compared to the average 23 years in the international market. Figure 5-10 depicts the age distribution of U.S.-flag tank ships scrapped between 1990 and 1995[16] and compares this distribution with the phaseout schedule of OPA 90.[17] In that period, less than 45 percent of the ships scrapped were 35 years old or less. Under the OPA 90 phaseout schedule, however, more than 80 percent of the U.S. fleet will be retired before age 35. A recent study by Clarkson (1996) shows that Jones Act tankers can ordinarily be expected to operate for an average of 35 years, with some smaller product tankers operating as long as 50 years (Figure 5-11).

Another key difference in domestic trade is a high degree of market uncertainty caused by factors such as competition from pipelines and imports, the 1996

[16]Figures 5-10 and 5-11 reflect retirement patterns of the total U.S.-flag tanker ship fleet, of which Jones Act vessels are the majority.

[17]Under OPA 90, Jones Act owners cannot adopt hydrostatically balanced loading (HBL) for the purpose of extending vessel life.

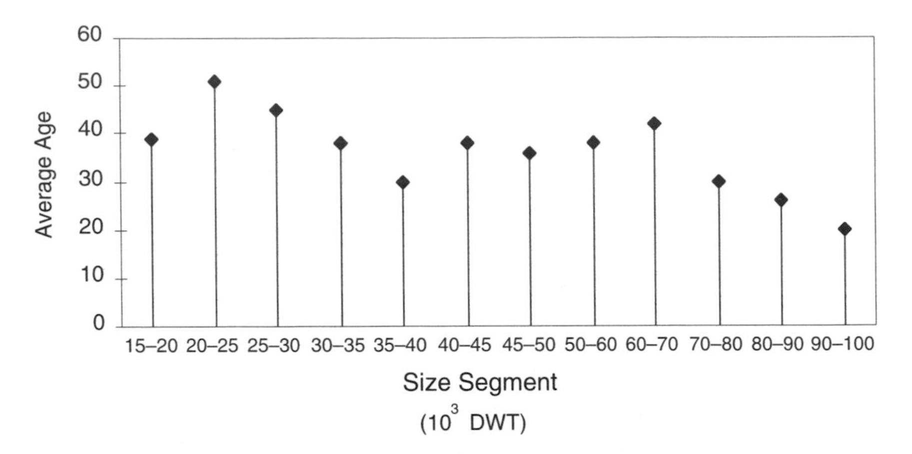

FIGURE 5-11 Average age of U.S.-flag tank ships when scrapped. Source: Clarkson Research, 1996.

decision by Congress to allow exports of Alaskan oil, and the large variation in projections of ANS production over the next 20 years.

Given all of the above, the difference in cost between continuing to operate a single-hull vessel until it is 35 years of age and purchasing a new double-hull replacement with retirement of the single-hull vessel at the age required by OPA 90 was calculated using the same methodology employed in Chapter 4.

Table 5-5 shows the "special survey break-even cost"[18] if the shipowner had the option of extending the life of the vessel. The average cost of a special survey is $2 million for a barge, $4 million for a 40,000 DWT tanker, and $6 million for a 120,000 DWT tanker. The MARAD 1996 report on product tankers estimates that the special survey for a product tanker could be as high as $6 million. Table 5-5 is based on the assumptions that the construction price of new Jones Act tankers would be 25 percent higher in U.S. yards than in foreign yards[19] and that, in the absence of OPA 90, all vessels would be retired at age 35.

The impact of Section 4115 on ship replacement, provided shipowners were willing to replace their tank vessels if that was the low-cost alternative, is the difference between the break-even special survey cost and the actual special survey cost. Details are shown in Table 5-6.

---

[18]The special survey break-even cost is the point at which the cost of the special survey in conjunction with continued vessel operation equals the cost of scrapping the existing vessel and replacing it with a new double-hull vessel.

[19]The estimate of 25 percent higher construction prices in U.S. yards is based on background data used in a study by ICF Kaiser (1995) and other general background information. There are recent indications that construction prices may be 50 percent or more higher in U.S. yards than in foreign yards. Higher construction price differentials would result in special survey break-even costs for Jones Act tank vessels increasing accordingly.

TABLE 5-5 Special Survey Break-Even Costs ($ million) for Jones Act Tank Vessels

	Capacity Reduction		
Tank Vessel Type	0 percent	10 percent	24 percent
Tankers < 50,000 DWT	10.9	4.0	65
Tankers > 50,000 DWT	18.0	6.0	63
Barges	4.7	2.0	137

Note: Although hydrostatically balanced loading does not extend the operating life of a vessel under OPA 90, it is shown for consistency with analyses of international regimes.

TABLE 5-6 Cost Impact of Early Retirement Due to Section 4115 on Jones Act Tank Vessel Fleet

Tank Vessel Type	Break-Even Cost (in $ million)	Estimated Special Survey Cost (in $ million)	Number of Special Surveys	Total Cost Impact (in $ million)
Tankers < 50,000 DWT	10.9	4.0	65	448.5
Tankers > 50,000 DWT	18.0	6.0	63	756.0
Barges	4.7	2.0	137	369.9
Total				1,574.4

The impact of Section 4115 on total cost is more than $1.5 billion. The key assumption here is that owners will decide to continue operations and to replace their single-hull vessels. However, if domestic market rates stay low and the market continues to contract, the committee's estimate of economic impact will be too high. Owners will simply opt to leave the market, regardless of regulations governing hull design. It is beyond the scope of the committee's charge to predict the future state of the Jones Act market regulations or related political actions that might be taken.

FINDINGS

Finding 1. The implementation of Section 4115 of OPA 90 will have a significant impact on the timing of vessel replacement in the domestic fleet. Historically, these vessels have remained in operation for 20 to 35 years or more. Under OPA 90, most of them will be retired well in advance of their traditional economic lifetime, with approximately 80 percent being phased out before they are 35 years old.

Finding 2. The requirement to replace vessels early could result in forgone utility costs to the domestic fleet of up to $1.5 billion if the entire single-hull fleet is phased out according to the OPA 90 schedule.

Finding 3. Shipyard capacity in the United States is adequate for new construction and conversion of Jones Act vessels. No bunching of orders is anticipated.

Finding 4. Although private capital is available and federal financing will facilitate economically viable double-hull projects, replacement of the Jones Act fleet will be discouraged if Alaskan oil production continues to decline and the higher freight rates needed to pay for double hulls cause domestic operators in the coastal products trade to lose business to pipelines and foreign tankers.

Finding 5. To the extent that domestic single-hull tank vessels are not replaced, there will be a corresponding loss in the transportation infrastructure, including shipyards, ship personnel, ancillary marine services, and suppliers.

REFERENCES

Bain, J.S. 1959. Industrial Organization. New York: John Wiley & Sons.

Clarkson Research, Ltd. 1990–1995. Tanker Register. London: Clarkson Research.

Clarkson Research, Ltd. 1996. Shipping Intelligence Weekly July 5.

Energy Information Administration (EIA).1996. Annual Energy Outlook 1996. Washington, D.C.: U.S. Department of Energy.

Gassman, W. 1996. Seasonality Trends. Study prepared for the Committee on Oil Pollution Act of 1990 (Section 4115) Implementation Review. Washington, D.C.

ICF Kaiser. 1995. Regulatory Assessment of Supplemental Notice of Proposed Rulemaking on Structural Measures of Existing Single-Hull Tankers. Prepared for U.S. Department of Transportation, Volpe National Transportation System Center, Cambridge, Massachusetts. July.

Maritime Administration (MARAD). 1995. Report on Survey of U.S. Shipbuilding and Repair Facilities. Washington, D.C.: U.S. Department of Transportation, December.

MARAD. 1996. Domestic Product Tanker Markets. Office of Statistical and Economic Analysis. Washington, D.C.: U.S. Department of Transportation, January.

Marine Strategies International (MSI). 1996. Oil & Tanker Market Update: New Lease on Life. London: Marine Strategies International.

Morgan, J.P. 1995. Overview of the financing outlook for the tanker industry. Prepared under the direction of James Hamilton (J.P. Morgan, Inc.) and distributed to the Committee on Oil Pollution Act of 1990 (Section 4115) Implementation Review. Washington, D.C., October 9.

Navigistics Consulting. 1996. Tanker Supply and Demand Analysis, Task 1 and Task 2. Study prepared for the Committee on Oil Pollution Act of 1990 (Section 4115) Implementation Review. Washington, D.C., June.

Newbold, J. L., and J. L. Grubbs. 1995. Presentation to the Committee on Oil Pollution Act of 1990 (Section 4115) Implementation Review. Washington, D.C. October 9.

National Research Council (NRC). 1991. Tanker Spills: Prevention by Design. Marine Board, Washington, D.C.: National Academy Press.

PIRA Energy Group. 1995. Forecast of Oil Demand. Prepared for the Committee on Oil Pollution Act of 1990 (Section 4115) Implementation Review. New York: PIRA Energy Group.

PIRA Energy Group. 1996. Energy Briefing. Presentation to Mr. Ran Hettena, New York, June 18.

State of Alaska. 1995. Revenue Sources Book, Department of Revenue. Anchorage: State of Alaska, Fall.

Wilson, Gillette & Co. 1994. Forecast Requirements for Tonnage in the Coastwise Petroleum Products and Specialty Trades 1990–2005. Prepared for the Maritime Administration, U.S. Department of Transportation. Washington, D.C.: Wilson, Gillette & Co.

6

Design, Construction, Operation, and Maintenance of Double-Hull Vessels

The performance of double-hull tank vessels with respect to such matters as structural integrity, safety, and prevention of oil spills in the event of accidents has been a subject of investigation for more that 20 years. This chapter begins with a discussion of the results of the committee-commissioned study of hull designs (see Appendix K),[1] notably oil outflow and ship stability characteristics. Next, the chapter reviews industry experience in design, construction, operation, and maintenance of double-hull tank vessels. The chapter concludes with a discussion of current design issues, including the need for revised design standards for double-hull vessels and the need for research on improved design tools. Additional information about research on double-hull vessel technology since 1990 is provided in Appendix L.

COMPARATIVE ANALYSIS OF DOUBLE-HULL AND SINGLE-HULL DESIGNS

Because few large double-hull tankers had been built before 1990, the promulgation of Oil Pollution Act of 1990 (P.L. 101-380) (OPA 90) and MARPOL Regulations 13F and 13G (MARPOL 13F and 13G) confronted naval architects with new design issues; existing national and international design regulations had been developed with single-hull tankers in mind. This new challenge stimulated creativity in the design process, as illustrated by the varied hull arrangements of

[1]The comparative study of double-hull and single-hull designs was performed under subcontract to the committee by Herbert Engineering Corporation, whose president is committee member Keith Michel.

the double-hull tankers constructed since the passage of OPA 90. Some of these designs, however, do not provide the high levels of environmental protection that can be achieved with double-hull vessels.

The comparative study of single-hull and double-hull vessels commissioned by the committee investigated the following:

- the effectiveness of hull design in reducing potential oil outflow following collisions and groundings
- ship stability as indicated by survivability characteristics after experiencing a collision and intact stability during load and discharge operations
- ship structural integrity as reflected by ballast condition, hull girder strength, and draft considerations when in ballast

Oil outflow, survivability, intact stability, ballast draft, and strength were evaluated for 27 tankers that either have been delivered or are under contract. Oil outflow and survivability calculations were also made for nine barges.

Double-Hull Tank Arrangements

The arrangement of tank vessel cargo tanks and ballast tanks has a major influence on a vessel's effectiveness in reducing oil outflow after an accident as well as its damage and intact stability. In particular, the subdivision of cargo and ballast tanks by centerline bulkheads can have important implications for oil outflow in the event of a collision or a grounding.

Box 6-1 shows the three most common cargo tank arrangements. Nearly all double-hull tankers exceeding 200,000 deadweight tons (DWT)—very large cargo carriers (VLCCs)—built to date have cargo tanks arranged three across. The cargo tanks on double-hull tankers of less than 160,000 DWT are usually arranged in "single-tank-across" or "two-tank-across" configurations. Approximately 60 percent of these vessels have single-tank-across cargo tanks in all or part of the cargo block.[2] All tankers exceeding 120,000 DWT delivered in the last three years have oiltight longitudinal bulkheads subdividing the cargo tanks. This is partly because of concerns regarding the outflow and stability characteristics of single-tank-across tankers and partly because of economic considerations. Suezmax tankers (about 150,000 DWT) that do not have oiltight centerline bulkheads require a large number of transverse bulkheads to satisfy MARPOL regulations for tank size and hypothetical outflow. As a result, construction costs for single-tank-across and two-tank-across double-hull tankers of approximately

[2]Data on cargo and ballast tank arrangements are from a compilation by Exxon Company International of configurations for 327 double-hull tankers comprising more than 95 percent of the world double-hull tanker fleet greater than 5,000 GT (gross tons). The compilation was derived from the Oil Companies International Marine Forum ship information questionnaires provided to Exxon by ship owners and from Exxon's internal inspection records.

BOX 6-1
Typical Cargo Tank Arrangements

The single-tank-across arrangement has a single center cargo tank and is frequently designed with additional structure to reduce sloshing when the cargo tanks are nearly full. The two-tank-across arrangement has a centerline bulkhead that results in port and starboard cargo tanks. Vessels of less than 160,000 DWT are typically arranged as a single tank across, two tanks across, or a combination thereof. For larger tankers, a minimum of three tanks across is required to satisfy MARPOL requirements on tank size and damage stability.

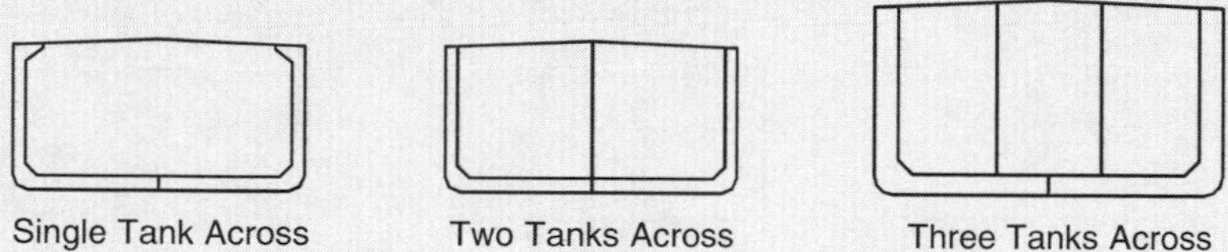

150,000 DWT are comparable. In contrast, many Aframax and Panamax tankers continue to be built with single-tank-across cargo tank arrangements. For tankers of less than 110,000 DWT, fewer transverse bulkheads are required within the cargo block, and the cost savings realized with the single-tank-across arrangement are more significant.

Box 6-2 shows typical ballast tank arrangements. The L tank is by far the most common configuration; it is found in 88 percent of double-hull tankers. Ten percent of the tankers have a combination of U, L, and S types, and 2 percent have a U design only.

Evaluating Oil Outflow

The International Maritime Organization (IMO, 1995) guidelines for evaluating alternatives to double-hull tankers were used to assess the relative oil outflow of different designs. Although intended for evaluating the outflow performance of alternative arrangements to the double-hull concept, these guidelines are also well suited to comparing the outflow performance of single-hull and double-hull tankers. The guidelines take a probabilistic approach based on historical data from collisions and groundings. (Other sources of oil spillage, such as explosions and operational discharges, are not included in the analysis.)

The IMO guidelines account for such factors as varying wing tank widths and double-bottom heights, internal tank subdivision, and the effects of tide.

BOX 6-2
Typical Ballast Tank Arrangements

The L tanks are the most commonly used configuration. These tanks are usually aligned with the cargo tanks although they occasionally extend longitudinally over two cargo tanks. The U tanks reduce asymmetrical flooding and are generally used when L tank arrangements fail to meet damage stability requirements. The U tanks extend over the full breadth of the ship and have a significantly greater free surface compared to a pair of L tanks. The S or side tanks are located entirely in the wing tanks. Because they do not extend into the double bottom, S tanks improve survivability when a vessel suffers bottom damage.

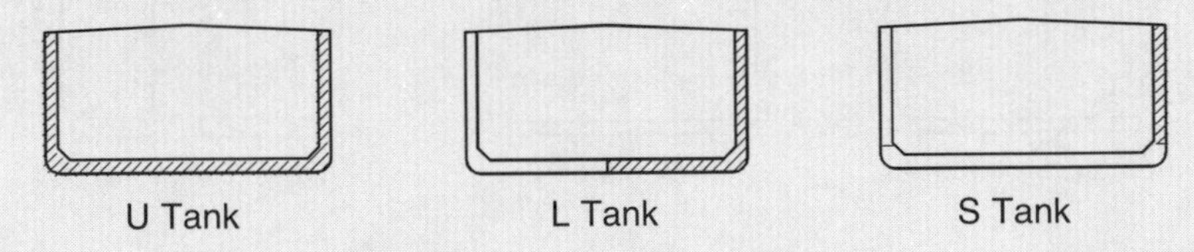

U Tank L Tank S Tank

Casualty statistics collected by classification societies were used to develop the expected distribution of side and bottom damage. The damage distribution functions were derived from about 60 tanker casualties involving primarily single-hull vessels. These distribution functions provide information on the expected penetration and the extent and location of damage from collisions and groundings.

In the case of a single-hull tanker, if the outer hull is penetrated adjacent to a cargo tank, the cargo tank will be breached and oil will flow out. For a double-hull tanker, outflow will occur only if the extent of penetration is sufficient to extend beyond the protective double-bottom or wing tanks, thereby piercing the inner hull and penetrating the cargo tank. The size of the spill is directly related to the number of cargo tanks breached and their size.

The likelihood that a double-hull tanker involved in a collision or grounding will spill oil is therefore largely influenced by the dimensions of the double-bottom and wing tanks. The amount of oil spillage is also impacted by the internal subdivision of the cargo tanks, which dictates tank sizes and the spacing of bulkheads forming tank boundaries. Naturally, larger cargo tanks will spill more oil. On the other hand, more closely spaced bulkheads increase the likelihood that more than one cargo tank will be damaged.

The quantitative results of the outflow analysis should be used with care because of the limited size of the casualty database, the nature of the incidents included,[3] and some simplifications in the calculation procedure. Nonetheless, the IMO methodology provides a rational basis for comparing tanker designs and, in the view of the committee, is currently the best readily available analytical approach.

Three outflow parameters were calculated: (1) the probability of zero outflow, (2) mean outflow, and (3) extreme outflow. (The mean and extreme outflow parameters measure volume rather than rate of outflow.) The probability of zero outflow is the likelihood that a collision or grounding will result in no oil spillage and indicates the effectiveness of a design in preventing oil spills. Mean (or expected) outflow is the weighted average of the cumulative oil outflow values for expected damage events and indicates the effectiveness of a design in mitigating the loss of oil due to collision or grounding. Extreme outflow is the weighted average of the cumulative outflow values for the most severe damage events and indicates the effectiveness of a design in reducing the number and size of large spills.

The outflow parameters for tankers evaluated in the comparative study are shown in Figures 6-1 through 6-4. Data points are plotted for each of the single-hull, double-sided, and double-hull tankers evaluated. To facilitate comparison, curves representing the least-squares fit of the single-hull and double-hull data are shown.

Zero Outflow

Figure 6-1 shows the probability of zero outflow values for tankers evaluated in the study. The calculations indicated that the probability of zero outflow is four to six times higher for double-hull tankers than for single-hull tankers. In other words, the projected number of spills for double-hull tankers is one-fourth to one-sixth the number of spills projected for single-hull tankers.

All cargo oil tanks on a double-hull tanker built to OPA 90 requirements are protected by ballast tanks or other nonoil spaces. Thus, many scenarios that would culminate in oil spillage from single-hull tankers do not result in penetration of the cargo tanks of a double-hull tanker. The probability of zero outflow is a function of the double-bottom and wing tank dimensions and is not affected by internal subdivision within the cargo tanks. In other words, centerline or other longitudinal bulkheads within the cargo spaces or ballast tanks have no influence on the probability of zero outflow.

---

[3]The committee recognizes that the probabilistic outflow methodology should ideally reflect the response of specific structural configurations. However, the same damage distributions are currently applied to both single-hull and double-hull vessels. This approach is likely to give conservative results (i.e., overestimates of outflow) when applied to double-hull designs, because recent studies have indicated that double-hull structures reduce the extent of damage from a collision or grounding. In certain cases the inner bottom or longitudinal bulkhead can withstand considerable deformation before being penetrated.

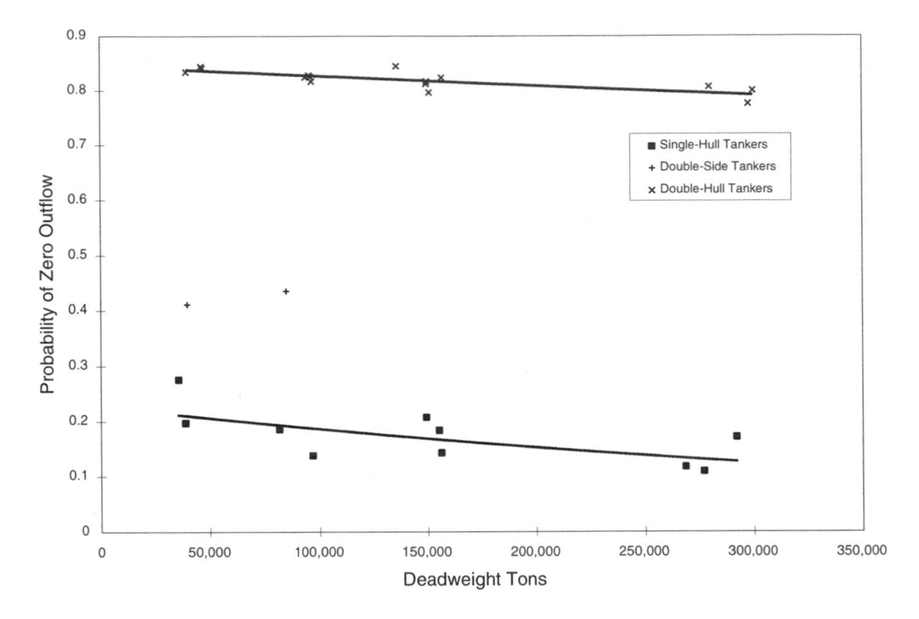

FIGURE 6-1 Probability of zero outflow for single-hull and double-hull tankers. Source: Herbert Engineering Corporation, 1996.

Mean Outflow

The mean outflow values for tankers evaluated in the comparative study are plotted in Figure 6-2. The mean outflow values for double-hull vessels are one-third to one-fourth the single-hull values, but mean outflow varies significantly even among double-hull tankers of the same size.

Mean outflow is influenced by the double-hull dimensions as well as the extent of internal subdivision. Wider wing tanks and deeper double bottoms tend to reduce the likelihood of a spill, thereby increasing the number of collisions and groundings with no spillage. Hence, an increase in wing tank width and double-bottom depth reduces the mean spill value. Greater internal subdivision also tends to reduce the quantity of oil spilled.

The variability in mean outflow values for double-hull tankers is primarily a result of differences in subdivision within the cargo block. Figure 6-3 is a plot of mean outflow, with tankers identified by the extent of longitudinal subdivision. Double-hull tankers without centerline bulkheads have approximately twice the expected outflow of designs with oiltight centerline bulkheads in way of all cargo tanks. Single-tank-across designs and designs with oiltight centerline bulkheads were found to have comparable outflow values when the vessel was subjected to bottom damage. However, single-tank-across designs performed less effectively when the vessel was subjected to side damage. The closer spacing of transverse

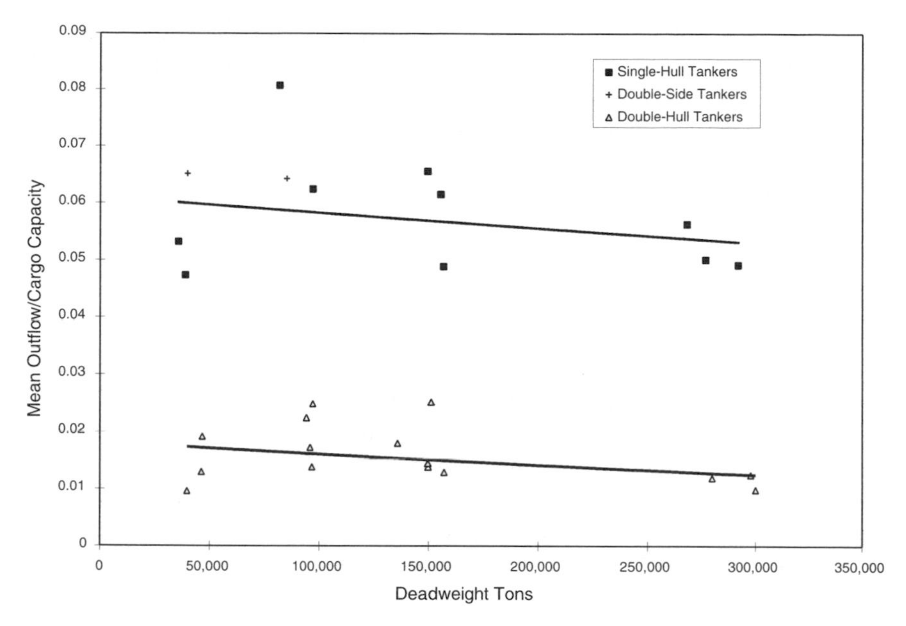

FIGURE 6-2 Mean outflow for single-hull and double-hull tankers. Source: Herbert Engineering Corporation, 1996.

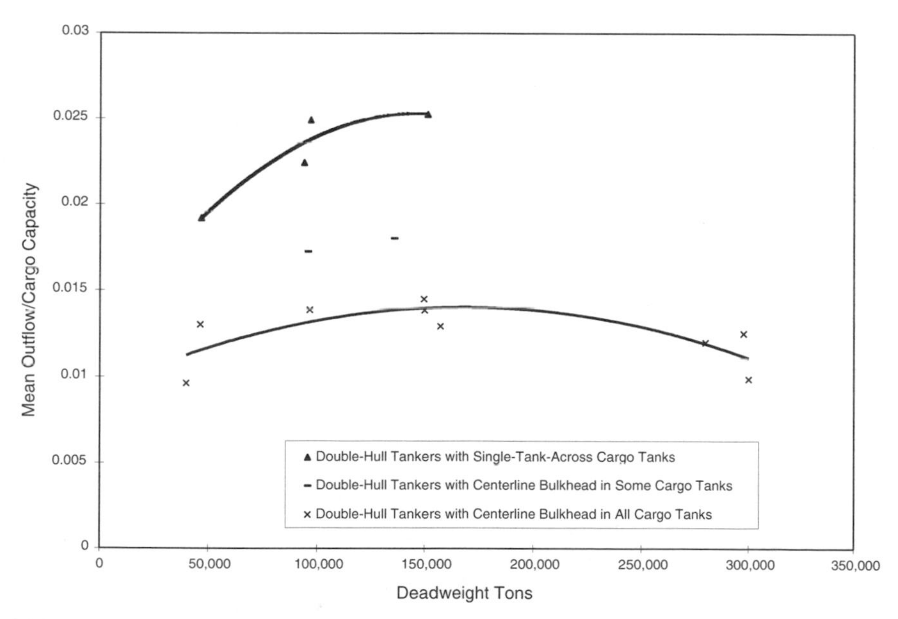

FIGURE 6-3 Variation in mean outflow with longitudinal subdivision for double-hull tankers. Source: Herbert Engineering Corporation, 1996.

bulkheads in these designs increases the probability that multiple cargo tanks will be breached. Once a single-tank-across cargo tank is breached, oil located all across the cargo compartment will flow out through the damaged side. Oil outflow is no longer limited to the oil being carried on one side of the vessel.

Extreme Outflow

Extreme outflow parameters are shown in Figure 6-4. There is considerable scatter in the data points, indicating that such characteristics as internal subdivision and draft-to-depth ratio have a significant impact on extreme outflow. Although the comparative analysis indicated that double hulls are very effective in reducing both the number of spills and the mean outflow values, their effectiveness in preventing large spills is less pronounced.

Double-hull vessels with single-tank-across arrangements perform more poorly with regard to extreme outflow than both pre-MARPOL and MARPOL 78 vessels of comparable size. Data points representing two double-side tankers and three double-hull tankers lie above the single-hull tanker trend line in Figure 6-4. All five of these double-hull or double-side designs have single-tank-across arrangements. Despite their poor performance relative to double-hull tankers with one or more longitudinal bulkheads in the cargo tanks, the three double-hull designs with single-tank-across arrangements meet all current U.S. and international regulations.

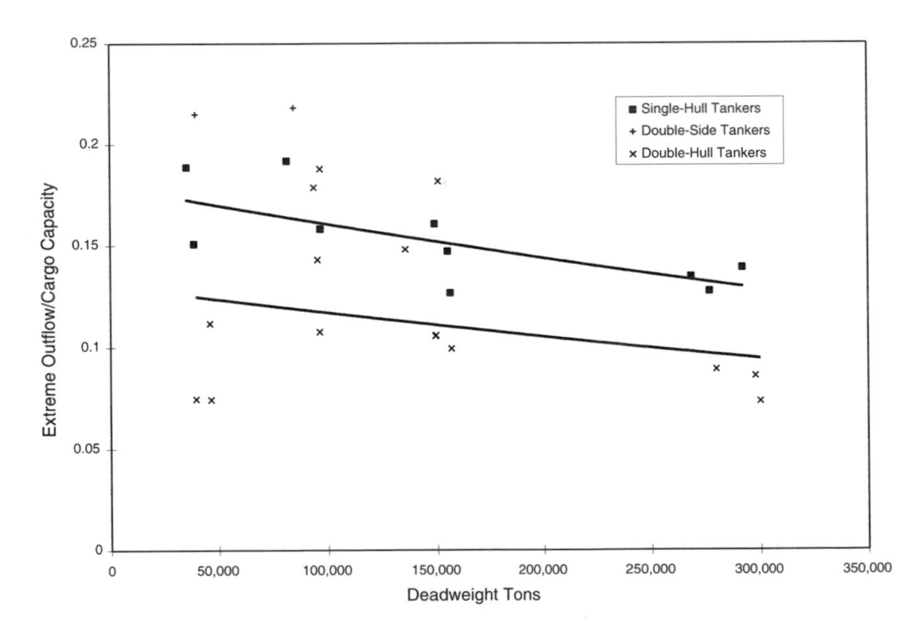

FIGURE 6-4 Extreme outflow for single-hull and double-hull tankers. Source: Herbert Engineering Corporation, 1996.

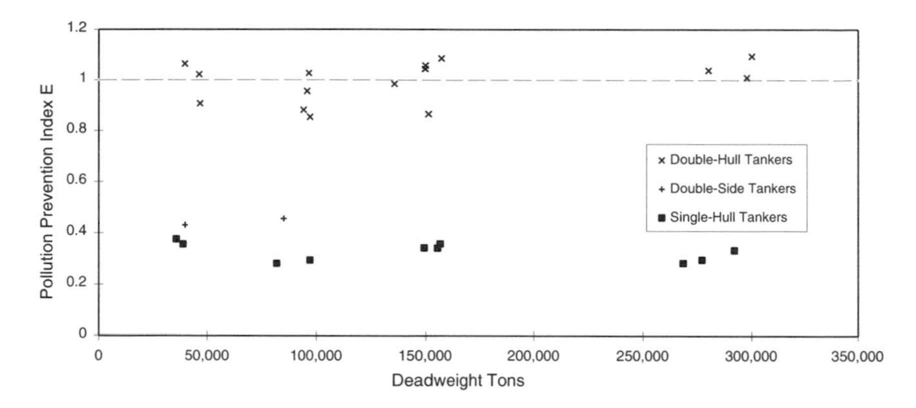

FIGURE 6-5 IMO pollution Index E for single-hull and double-hull tankers. Source: Herbert Engineering Corporation, 1996.

Combined Outflow Performance

The IMO Pollution Prevention Index E provides an overall picture of outflow performance. The three outflow parameters for a given design are combined, using weighting factors, and then compared to the outflow parameters for an IMO reference ship of similar size.[4] An Index E greater than or equal to 1.0 indicates equivalency to IMO's reference designs.

Figure 6-5 shows the Index E for single-hull and double-hull tankers evaluated in the comparative study. Single-hull tanker values generally fall between 0.3 and 0.4, whereas double-hull tanker values lie between 0.9 and 1.1. Sixty percent (9 of 15) of the double-hull designs have indices greater than 1.0, indicating equivalency to IMO reference ships. In general, ships with longitudinal oiltight bulkheads in the cargo holds have the highest indices.

Outflow Performance of Tank Barges

The probability of zero outflow and the mean outflow values for tank barges evaluated in the comparative study are plotted in Figures 6-6 and 6-7, respectively. The results are similar to those for tankers. Double-hull tank barges exhibit substantial superiority in both probability of zero outflow and mean outflow when compared to single-hull tank barges.[5]

---

[4]Sketches of the IMO reference ships are provided in Appendix K. These reference designs do not incorporate the minimum subdivision acceptable under current MARPOL regulations. They were selected because they exhibit a favorable oil outflow performance.

[5]Although tankers in the 5,000 to 25,000 DWT range were not evaluated in this study, the committee believes that the outflow values for double-hull barges would be slightly better than for double-hull tankers. This observation is based on the relatively low freeboards (distances from sea surface to deck) of barge designs, which tend to reduce the outflow from bottom damage.

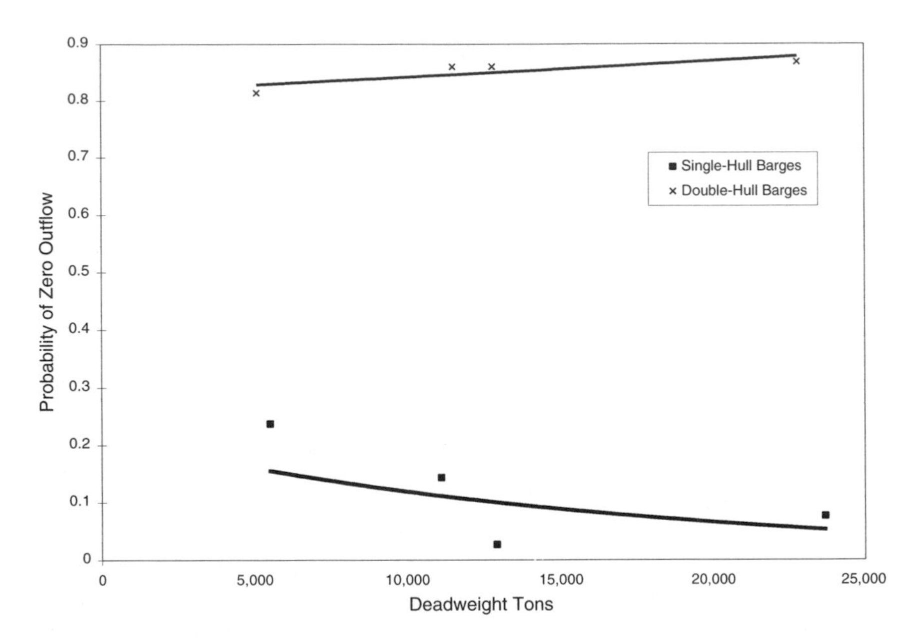

FIGURE 6-6 Probability of zero outflow for single-hull and double-hull tank barges. Source: Herbert Engineering Corporation, 1996.

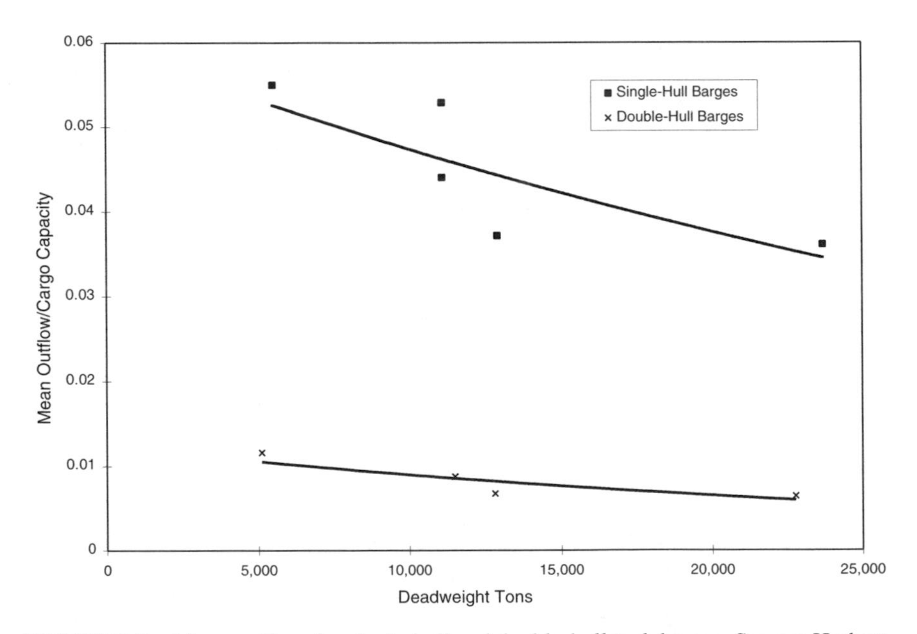

FIGURE 6-7 Mean outflow for single-hull and double-hull tank barges. Source: Herbert Engineering Corporation, 1996.

TABLE 6-1 Survivability Indices for Single-Hull and Double-Hull Tankers

	Survivability Index (%)	
Vessel Capacity (DWT)	Single Hull	Double Hull
35,000–50,000	95.0	95.9
80,000–100,000	99.6	99.7
135,000–160,000	100.0	99.7
265,000–300,000	99.9	100.0

It is important to remember that this study investigated the relative performance of different designs to mitigate outflow if they experienced a collision or grounding that breached the outer hull. The overall outflow performance must also take account of the likelihood that a given vessel will experience such an accident. Therefore, a comparison of barges and tankers cannot be made on the basis of outflow parameters alone.

Ship Stability

Survivability

Survivability is a measure of a vessel's ability to survive (i.e., not capsize or sink) after sustaining damage to the hull. A probabilistic methodology was used to assess the survivability of single-hull and double-hull tankers. Probabilistic density distribution functions for side damage, as contained in the IMO guidelines,[6] were used to determine all possible damage events and their probability.

The average survivability indices for each tanker size are given in Table 6-1. There is no discernible difference between the survivability characteristics of single-hull and double-hull tankers. The survivability indices generally exceeded 99 percent for tankers of more than 80,000 DWT. However, two of the ships in the 35,000 to 50,000 DWT range had values of 87.2 percent and 92.5 percent, respectively. These values were more heavily influenced by the degree of compartmentation within the engine room and adjacent spaces than by differences between single-hull and double-hull arrangements.

Intact Stability

Single-hull tankers are generally stable under all loading conditions. Therefore, stability during operation when no damage has occurred (intact stability)

[6]Assessment of survival or nonsurvival is based on damage survival requirements defined in Regulation 25 of Annex I of MARPOL. The index of survivability is developed by summing the probabilities for all damage cases that survive the damage criterion. A survivability index of 97 percent indicates that a vessel is expected to survive collisions breaching the outer hull 97 percent of the time.

has not been a concern of tanker operators in the past. In contrast, certain double-hull tankers can become unstable during cargo and ballast operations. The reduction in the intact stability of double-hull tankers compared to single-hull tankers is due to the increased height of the center of gravity of a double-hull vessel (because the double bottom raises the center of gravity of the cargo) and to the large free surface effects[7] that some ships with double-hull designs experience during cargo and ballast operations. The magnitude of the free surface effect depends on tank arrangement and is discussed further in Appendix K.

As expected, all of the single-hull tankers evaluated in the comparative study were inherently stable. Of the double-hull vessels, only 73 percent (11 of 15) were inherently stable. The four designs with a potential for instability all had single-tank-across arrangements. None of the vessels with these designs, however, capsized when subjected to the worst-case loading scenario. Three have maximum angles of loll of less than 8 degrees.[8] The load restrictions required to ensure positive stability for these vessels are quite straightforward—namely, the monitoring of any two ballast tanks. With all ballast tanks 2 percent full, these designs maintain positive stability through all possible cargo load conditions. The fourth design has a potential angle of loll of 15 degrees. With all ballast tanks 2 percent full, the vessel can be unstable when cargo tanks are partly full. The load restrictions required to assure positive stability for this vessel necessitate the monitoring of both ballast and cargo tanks.

Recent studies by the Society of Naval Architects and Marine Engineers (SNAME), the American Bureau of Shipping (ABS), and the U.S. Coast Guard indicate that it is possible to design inherently stable double-hull tankers (Goodwin et al., 1996; Moore et al., 1996). However, no risk or cost-benefit analyses have been undertaken to assess the costs of inherently stable designs and their impact on ship safety and marine pollution.

Evaluation of Ballast Conditions

Minimum dimensions for wing tanks and double-bottom tanks are specified in MARPOL 13F. The committee's review of recent tanker construction showed that most double-hull tankers are designed with double-bottom and wing tank dimensions in excess of the minimum requirements. This is probably the result of a desire to provide better access to ballast tanks for inspection and construction purposes, owner requirements for deeper ballast drafts than the IMO minimum

[7]Free surface effects are cargo or ballast movements that occur in partially loaded tanks as a vessel heels. Free surface effects tend to reduce stability.

[8]If an initially upright ship is unstable, it will tend to heel to one side. As the ship heels, the buoyancy force exerted on the ship's hull shifts outboard, counteracting this tendency to heel. The ship will come to rest at an angle of loll when the buoyancy-induced righting moment equals the heeling moment. If the righting moment is not sufficient to counteract the heeling moment, the vessel will capsize.

values, and considerations regarding structural strength and oil outflow. All double-hull designs evaluated in the comparative study had forward drafts at least 19 percent deeper than the IMO minimum, and most designs had drafts that were more than 50 percent greater than the IMO minimum.

The 1992 Amendments to MARPOL specified that wing tanks and double-bottom tanks used to meet the ballast draft requirements "shall be located as uniformly as practicable along the cargo tank length." This requirement leads to a generally homogeneous longitudinal distribution of ballast. As a result, double-hull tankers typically have higher hogging moments[9] in the ballast condition than most MARPOL single-hull tankers, whose ballast is concentrated closer to amidships. Most of the double-hull designs evaluated in the comparative study had still-water bending moments in the ballast condition approaching the maximum permissible value assigned by the classification society. It is difficult to say what impact this may have on a ship's structure because fatigue is influenced primarily by cyclic loading. High still-water bending moments remain a matter of concern to some shipowners.

EXPERIENCE WITH DOUBLE-HULL TANK VESSELS

The committee developed two questionnaires (see Appendix C) to elicit information from industry representatives on the design, construction, operation, and maintenance of double-hull tank vessels. One questionnaire was sent to shipyards, classification societies, and naval architects to obtain information on double-hull tankers designed or built since 1990. This questionnaire covered ship design characteristics, producibility[10] differences between single-hull and double-hull construction, and maintenance-related design problems and solutions. The 21 responses received provided information on the design characteristics of 91 ships with 61 different designs that either had been delivered or were scheduled to be delivered between 1991 and 1996 (see Appendix M).[11] These vessels comprise about 28 percent of the world's double-hull tanker fleet exceeding 20,000 DWT.

A separate questionnaire was sent to owners and operators of double-hull tankers and to shipyards that build double-hull tankers, requesting information on their experience in operating, maintaining, and inspecting double-hull vessels.

---

[9]Bending moments are induced in the ship's hull girder when the distribution of downward forces exerted by the weight of a vessel differs from the distribution of the upward or buoyancy forces exerted by seawater. If the weights are concentrated toward the ends of the vessel and the buoyancy toward the middle, the vessel hull will tend to deflect downward at its ends. This is called a "hogging" condition.

[10]Producibility relates to the ease of fabrication.

[11]The number of vessels in different size ranges was as follows: 21 vessels of less than 60,000 DWT; 22 vessels of 60,000 to 100,000 DWT; 27 vessels of 100,000 to 200,000 DWT; and 23 VLCCs (200,000–350,000 DWT).

TABLE 6-2 Advantages and Disadvantages of Double-Hull Compared to Single-Hull Tankers

	Advantages	Disadvantages
Cargo operations	Faster cargo discharge and good cargo out-turn[a] Easier and faster cleaning of cargo tanks	
Construction and producibility		Higher cost
Inspection and maintenance		Higher maintenance cost Need for continuous monitoring and maintenance of ballast tank coatings
Operational safety		Structural safety concerns, intact stability, and increased still-water bending moment Difficult access to and ventilation of ballast spaces

[a]Out-turn is defined as the ratio of the quantity of cargo discharged to the quantity of cargo loaded.

The survey was limited to crude oil, oil product, chemical, and oil-bulk-ore (OBO) carrier operators.[12] Responses were received from 14 owners or operators of double-hull tankers or OBO carriers. Data were obtained on 91 double-hull tankers (33 chemical or product carriers, 25 OBO carriers, and 33 crude oil carriers),[13] as well as 42 chemical carriers of between 8,000 and 40,000 DWT with a double bottom, double sides, or both, and on 3 double-hulled barges (see Appendix M).

According to the operators surveyed—most of whom recognize the double-hull tanker as an industry standard—there are significant differences between the operation of double-hull and single-hull tankers. The reported advantages and disadvantages of a double-hull compared to a single-hull tanker are summarized in Table 6-2. The general opinion was that double-hull tankers can be operated safely, albeit with increased resources. However, some operators expressed concern that procedures for the safe operation of double-hull tankers are left to the

[12]The committee acknowledges that a survey of operational experience on liquid natural gas, liquid propane gas, and ammonia carriers could provide some useful information, but such a survey would require an extensive evaluation of the operational differences between the different types of vessels before any comparisons could be made with oil tankers.

[13]The number of vessels in different size ranges was as follows: 36 vessels of less than 60,000 DWT; 11 vessels of 60,000 to 100,000 DWT; 34 vessels of 100,000 to 200,000 DWT; and 10 VLCCs (200,000–350,000 DWT).

discretion of the vessel owner rather than being mandated through regulations or classification society rules.

Cargo Operations

Almost all of the operators identified the major advantage of double-hull tankers as faster discharge of cargo and a reduction in the amount of cargo remaining on board following discharge. Cleaning of cargo tanks is also easier and faster on double-hull tankers. These advantages accrue because most of the ship's framing is located in the ballast tanks, leaving smooth surfaces in the cargo tanks. Less crude oil washing is required in tanks where most of the surfaces are smooth. In addition, sumps in the double bottom allow most of the stripping (final stages of cargo discharge) to be done by main cargo pumps rather than with lower capacity stripping pumps as in single-hull tankers, thereby saving time and reducing the volume of residual cargo that cannot be pumped out.

Construction and Producibility

Comparative construction and cost data for single-hull and double-hull vessels were obtained from 13 shipyards that responded to the committee's questionnaire. Comparative labor hours or cost data were obtained on the weight and type of steel used, coatings, machinery and outfitting, and construction time. Table 6-3 shows the differences associated with double-hull tankers compared to single-hull designs.

Steel Use

A double-hull structure requires more steel than a single-hull structure, and the fabrication time is longer. Early pre-MARPOL ships were built of mild steel

TABLE 6-3 Comparison of Producibility Factors for Double-Hull and Single-Hull Tankers

	Percentage Change, Double-Hull Compared to Single-Hull Tankers		
Producibility Factor	60,000–100,000 DWT	100,000–200,000 DWT	200,000–350,000 DWT
Steel weight	+10	+10 to +13	+10 to +20
Steel fabrication (man hours)	+5 to +35	+10 to +20	+10 to +38
Coatings	+20 to +79	+35 to +70	+35 to +90
Machinery and outfitting	0 to +15	0 to +10	0 to +14
Construction time	0 to +16	0 to +18	+15 to +31 (+1 to +2 months)

with a yield strength of 24 kg/mm^2. When tanker size increased in the early 1970s, high tensile steels (HTS) with yield strengths of 32 or 36 kg/mm^2 were increasingly used, both to reduce construction costs and to enable ships to carry more cargo deadweight for the same displacement. Experience has shown that the higher operational stresses associated with HTS increase the risk of fatigue cracks developing. The details of structural workmanship and alignment of strength members affect the initiation of fatigue cracks. Localized stress concentrations in the complex double-hull structure can result in fatigue cracks, which range from nuisance cracks to cracks severe enough to cause leaks or loss of structural strength. Cracks on surfaces between the cargo and the ballast tanks can allow oil to leak into ballast spaces and produce hydrocarbon vapors that pose a risk of explosion or fire.

There were instances of cracking in some VLCCs in the late 1980s, leading to a subsequent reduction in the percentage of HTS used in large tankers. Design data obtained in response to the committee's survey indicated that HTS is used extensively in the hulls of double-hull tankers between 60,000 and 350,000 DWT. There is a tendency, however, to reduce the use of HTS with a yield strength of 36 kg/mm^2 in favor of steel with a yield strength of 32 kg/mm^2 in newer double-hull VLCCs, presumably in response to concerns about fatigue cracking.

The newest (post-1990) double-hull tankers have shown no significant structural problems. It is not clear whether this good structural record is a result of the young age of most double-hull tankers or of special attention paid to design detail. Recognizing the importance of design detail in ensuring structural integrity, classification societies have included fatigue evaluation in their approval procedures. However, some owners consider the structural requirements in classification society rules to be insufficient and carry out structural analyses that exceed the requirements of the classification society. The Tanker Structure Cooperative Forum (TSCF), which is composed of classification societies, oil companies, and independent tanker operators, has developed guidelines that provide information on critical areas in double-hull tankers and include a historical catalogue of failures in structural details.

Coatings

The ballast tanks of double-hull tankers are coated to protect the steel against corrosion that can accelerate fatigue cracking. Coating quality and maintenance are particularly important on HTS structures, which tend to be thinner than corresponding mild steel structures. The most common type of coating is coal tar epoxy applied in two coats that produce a total dry film thickness of 250–300 microns.[14] Some companies use light-colored coatings, such as bleached coal tar epoxy, to facilitate visual inspection of the tanks. Others favor pure coal tar epoxy.

[14] 1 micron = 10^{-6} m.

There are no reliable data on the life expectancy of various types of coatings. Most double-hull tankers have been in operation for a relatively short time (less than five years), and no major corrosion problems have been reported on these vessels. Coating failures on relatively new vessels have been attributed to poor workmanship during construction. Some owners of 15- to 20-year-old double-hull tanker fleets have reported significant corrosion in double-hull spaces. Corrosion problems have led to major steel replacement in some double-hull vessels and have contributed to the scrapping of a number of double-hull vessels.

Some tanker operators have detected "microbial-influenced corrosion" in the uncoated bottom plating of cargo tanks. Microbes present in the settled water and sludge in cargo tanks are believed to cause accelerated rates of corrosion.[15] Although microbes have been present in tank sludge in the past, there is speculation that the higher temperature and other conditions found in some double-hull oil tankers foster explosive growth of the microbe population (Marine Log, 1996). This phenomenon is being studied, but its impact is not yet clear (Huang, 1996).

Shipowners have made estimates of coating life. Many expect coatings to last throughout the life of the vessel; others expect coatings to last 10 to 15 years. All survey respondents emphasized the importance of surface preparation and adequate coating thickness in prolonging coating life. Coating replacement is expensive, and a replaced coating is generally less durable than the original. Some owners and operators believe that mandatory requirements for coating application are necessary to prevent owners from specifying low-quality coatings at the time of construction. The classification societies do not inspect coatings to verify application, and it is the owner's responsibility to see that a coating is applied in accordance with the manufacturer's recommendations.

Construction Time

Survey respondents indicated that the construction time for a double-hull tanker can be greater than that for an equivalent single-hull vessel, depending on ship size. The increase in construction time is related in part to the increased amount of steel needed for the more complex double-hull structure and the increased coating area. As a result of the longer construction time, shipyards, classification societies, and marine architects expect double-hull tankers to be 7 to 15 percent more expensive than equivalent single-hull designs. The committee's economic analysis indicated an increase in capital costs for double-hull tankers of between 9 and 17 percent.

Problems encountered in double-hull construction include difficulties in painting narrow double-hull spaces, in providing adequate ventilation of double-hull spaces, and in providing adequate access to double-bottom ballast tanks when the side tanks and double bottom are divided to provide damage stability. It is

---

[15]Corrosion rates of 0.55 to 1.00 mm per year have been reported (Marine Log, 1996).

anticipated that as shipyards build more double-hull vessels, there will be a reduction in labor hours as a result of optimization of the structural arrangements and details and more extensive use of computer-aided manufacturing processes. The slotting of longitudinals through webs and bulkheads, for example, occurs thousands of times during construction of a double-hull tanker. Developing more producible details that lend themselves to robotic welding is expected to yield substantial savings. Recent developments in this area have been reported by Odense Steel Shipyard in Denmark and Kawasaki Heavy Industries, Ltd., in Japan (Motoi et al., 1995; Tang-Jensen, 1995).

Inspection and Maintenance

Survey respondents said that frequent inspection and maintenance of coatings were essential for safe and economic operation of double-hull tankers and were also more critical than for single-hull vessels. Apart from this increased inspection requirement, inspection and maintenance practices do not differ significantly between single-hull and double-hull tankers. The structural and coating survey requirements for both single-hull and double-hull tankers have increased since the passage of OPA 90 and MARPOL 13G. In response to these regulations, the U.S. Coast Guard and International Association of Classification Societies (IACS) have established survey requirements. The main financial and operational impact of the hull portion of the Enhanced Survey Program stems from the requirement for a dry-dock survey to complete the special hull survey, provision of access for close-up surveys, periodic evaluation of coatings and increased monitoring of tanks with poor coatings, and thickness measurements during special and intermediate surveys.[16]

The opinion among those responding to the committee's survey was that more resources are required to maintain double-hull than single-hull vessels. Structural and coating inspections required by an owner are usually performed by the ship's crew. Inspection frequencies for each tank vary from every couple of months to once a year. In addition, some companies have inspections conducted by independent surveyors at two- to five-year intervals. Many of the companies monitor ballast tanks on double-hull tankers more closely and more frequently than those on single-hull tankers. If coating failure is detected, minor repairs are often carried out during inspections.

[16]The IACS Enhanced Survey Program includes close-up surveys (visual inspection of the structure at close range), thickness measurements of areas subject to corrosion, and evaluation of the tank coating system. If the surveyor finds the tank coating to be in poor condition, the tank will be inspected annually.

Operational Safety

Intact Stability

Intact stability, which has not been a major concern for single-hull tank operators, has become a concern in the operation of some double-hull tankers, as noted in the committee's comparative study discussed earlier.

Access Hazards and Explosion Risk in Ballast Spaces

The structural arrangement of the double-side and double-bottom tanks of double-hull tankers is usually cellular. Safe access to these spaces is essential to monitor ballast tanks, conduct surveys required by classification societies, and maintain ballast piping. In addition, access may be needed to rescue an injured person from a double bottom in the event of an accident.

Opinions varied regarding the accessibility of ballast tanks. Some operators considered access on double-hull vessels to be easier than on single-hull vessels and saw no need for regulations on access. Others reported that access to double-hull spaces was difficult, escape distances in an emergency were long, and design complexity required ship personnel to have good knowledge of tank configuration before entering. These respondents favored mandatory access standards.[17] All respondents emphasized the need for adherence to stringent safety procedures by workers when entering tanks.

Double-hull designs that have taken accessibility into account have horizontal stringers (or decks) and large longitudinal stiffeners or built-in walkways in side tanks to provide access. Large openings are used in intermediate decks for direct access to other levels; these openings can be a safety hazard unless railings are installed to prevent falls. Some designs have separate rescue hatches in every tank that allow direct access to the main deck in an emergency, and in double-bottom tanks the number of bays without direct access has been minimized.

Questions have been raised about the ventilation of cellular double-hull spaces, and the opinions of operators on this issue varied greatly. Some believed that risks associated with a possible lack of adequate ventilation have been over emphasized. Others thought that even after forced ventilation, these spaces might contain pockets lacking oxygen or, in the case of oil leakage, pockets of flammable gases that could cause fires and explosions. The latter group urged installation of

---

[17]IMO has some general requirements for accessibility, inerting, venting, and gas-freeing of cargo and double-hull spaces. These requirements include considerations governing safe access, the capability to supply fresh air to double-hull spaces, the capability to inert double-hull spaces, and the monitoring of other vapor concentrations. These requirements are included in the International Convention on Safety of Life at Sea (SOLAS), Chapter II-1, Regulation 12-2, and Chapter II-2, Regulation 59.

fixed gas detection systems in the tanks, although some noted that instrumentation could lead to complacency.

Many operators rely on portable fans and hatch openings for ventilation. According to TSCF guidelines, this is not effective for all ballast tanks. Some operators have had dedicated purge pipes installed in the tanks or use ballast pipes to provide air circulation in ballast spaces.[18]

Many operators consider it important to have the capability of making all ballast and void spaces inert (i.e., reducing oxygen content) by installing emergency connections from the inert gas system to ballast pipes. Others believe that fixed inert gas systems for ballast spaces should be a requirement. The use of ballast pipes to make a tank inert raises concern because inert gas can migrate to other ballast tanks through ballast pipes. Although some owners and operators expressed concern about the lack of mandatory requirements, the respondents generally considered existing guidelines for access, ventilation, and inerting of ballast spaces as defined in TSCF guidelines and the International Safety Guide for Oil Tankers and Terminals to be adequate.

Structural Issues

Although the structural design concept of a double-hull tanker is essentially the same as that for a single-hull tanker, the structural response of a double-hull tanker exhibits some distinctive features that may warrant special consideration. In particular, a double hull tends to be stiffer than its single-hull counterpart, and this can affect residual stresses induced during construction and local stresses induced by operational loads, both of which may initiate fatigue cracks. There are also concerns over high still-water bending moments, although their impact on ship structure is unclear.

The strength design criteria typically employed for tanker hull structures are those set forth in classification society rules. Until very recently, the rules of most societies were semiempirical, experience-based standards that reflected the experience of ships at sea but could not readily be extrapolated to new designs and new technology. These traditional rules were based on yielding as the primary structural failure mode, with other failure modes (such as buckling and fatigue) accounted for by safety margins in the criteria. It is generally recognized that a more rational and consistent method of establishing criteria is necessary to take account of the dominant structural failure modes of yielding, buckling, and fatigue. Since 1990, some classification societies have therefore taken steps to improve their structural standards by developing new hull structural strength criteria specifically for double-hull tankers and other ship types (see, for example, Chen et al., 1993).

---

[18]Ballast pipes cannot be used for ventilation during ballast transfer operations.

Salvage

The committee was not able to find any new information on the performance of double-hull tankers in salvage situations. Thus, the issues raised in an earlier study on double-hull tankers remain unresolved (NRC, 1991).

The salvors interviewed by the committee commented that the large number of ballast tanks in double-hull tankers provides flexibility in a salvage situation. Therefore, salvage of a double-hull tanker may be easier than salvage of a single-hull tanker. However, salvage procedures—and the associated benefits and disadvantages of double-hull spaces—are highly dependent on actual circumstances.

Successful salvage of a damaged tanker is critical in minimizing oil outflow. Although the primary emphasis of the tanker industry is on accident prevention, at least one of the tanker companies interviewed has accorded high priority to the salvage capabilities of its tankers. Vessels have been equipped to facilitate salvage operations and have crews trained for salvage situations.

DESIGN OF DOUBLE-HULL TANK VESSELS

Design Standards

The regulations governing tanker design were developed primarily with single-hull vessels in mind, although the stability and strength characteristics of double-hull vessels are quite different from those of the traditional single-hull tanker. Existing and proposed regulations pertaining to oil outflows, intact stability, and survivability of double-hull tankers are summarized in Table 6-4.

As demonstrated by the comparative study described in Appendix K, present regulations do not ensure consistently high levels of environmental performance by double-hull tankers. Where practical, IMO is committed to replacing the current deterministic regulations[19] with probabilistic-based regulations. Work is under way at IMO to develop a performance-based regulation for evaluating tanker outflow. IMO is also harmonizing its damage stability criteria for all types of ships based on a probabilistic methodology that will eventually include tank vessels and chemical carriers.

Performance-based criteria establish a minimum level of performance but do not specify the means of attaining this minimum. Such criteria generally take a probabilistic approach, so that the influence of a given incident on overall design is proportional to its likelihood of occurrence and to its severity or repercussions.

[19]An example of a deterministic criterion is the IMO raking bottom damage regulation. This is a damage stability criterion that assumes extensive damage to the bottom shell while the double bottom remains intact. For tankers greater than 75,000 DWT, damage is assumed to begin at the bow and extend aft over 60 percent of the vessel's length. This type of criterion encourages designers to place bulkheads immediately beyond the specified damage extent but does not necessarily lead to optimum designs.

TABLE 6-4 Existing and Proposed Regulations Relating to Oil Outflow, Intact Stability, and Survivability Performance of Double-Hull Tankers

Regulation	Requirements, Scope, Status
Oil outflow from collisions and groundings	
MARPOL 13F	Establishes minimum dimensions for wing and double-bottom tanks comprising outer hull Consistent with USCG requirements established in response to OPA 90
Regulations 22–24, Annex I to MARPOL 73/78	Define hypothetical outflow and tank length requirements governing extent of cargo tank subdivision Regulations 22–24 being revised in light of probabilistic methodology for oil outflow analysis
Intact stability of tankers	
None at present	Intact stability to meet criteria recommended by IMO (Resolution A.749(18), 3.1.2.10)[a] normally exceeded by double-hull tankers through design Two possible approaches: (1) through design only, and (2) through combination of design and operational procedures Maritime Safety Committee of IMO addressing issue of intact stability for double-hull tankers. MARPOL Draft Regulation I/25A calls for assurance of positive intact stability, both in port and at sea, through design only[b]
Survivability of tankers	
Regulation 25, Annex I to MARPOL 73/78	Specifies extent of damage tanker must be able to survive
MARPOL 13F	Defines raking bottom damage criterion that supplements Regulation 25 Damage stability criteria for all types of ships being harmonized by IMO based on probabilistic methodology

[a]IMO code on intact stability for all types of ships is covered by IMO instruments.

[b]The Marine Environment Protection Committee of IMO will circulate Draft Regulation I/25A with a view toward adoption in September 1997.

For example, current regulations specify minimum wing tank and double-bottom clearances. A performance-based criterion might establish a minimum value for "probability of zero outflow." Rather than a uniform double hull, a more effective design might have a deeper double bottom located below the forward cargo tank and narrower wing tanks located outboard of the aft cargo tanks.

Performance-based criteria are more difficult to develop than the traditional deterministic criteria and are generally more complicated in their application. An assessment is required of both the relative probability of each possible event and the associated risks to the vessel's safety and to the marine environment. Thus, the costs and benefits of ship safety and spill mitigation measures must be understood before effective performance-based criteria can be developed.

Nonetheless, properly developed performance-based criteria have many advantages. They give the designer the freedom to optimize a design for minimum construction cost while ensuring that safety and environmental performance standards are met. They are also more adaptable to new concepts. For instance, a performance-based probabilistic outflow criterion would have predicted the poor outflow performance of many of the single-tank-across double-hull tankers. The methodology used to develop performance-based criteria is independent of the required index or performance level, thereby allowing the required level of vessel performance to be readily revised in the light of experience or in response to changes in cost-benefit scenarios.

Progress in Design

In the years after the promulgation of OPA 90, coordinated research on the performance of double-hull tankers has been pursued at several centers in the United States, Japan, the Netherlands, Denmark, and Norway (see Appendix L). Structural research is proving beneficial in providing improved design tools to incorporate fatigue and structural performance in accident scenarios into double-hull tanker designs.

Advances in finite element stress analysis techniques have made it possible to determine accurately the detailed local stresses due to operational loads, which can result in initiation of fatigue cracks in double-hull vessels. For the most part, analyses of this type are now carried out routinely as an integral part of the design process, and the application of fracture mechanics has already paid dividends in improving the fatigue life of ship hulls. Nevertheless, there is potential for further progress.

Since V.U. Minorsky's efforts in the late 1950s to correlate the interpenetration of colliding ships using accident data (Minorsky, 1959), there has been continuing research aimed at accounting more accurately for the structural details and approach characteristics of colliding ships. Collision analysis has been greatly aided by modern nonlinear finite element methods, which are now being used to optimize double-hull designs with respect to plate thickness, steel strength, and the positioning of inner and outer hull plates, side stringers, and transverse webs. Verification of analytical procedures by means of scale-model tests and actual collision data, where available, is a necessary part of the approach. Some full-scale collision tests have been conducted (Vredeveldt and Wevers, 1992; Vredeveldt and Wevers, 1995; Wevers et al., 1994).

Most of the characteristics of structural failure in tanker grounding incidents can be analyzed with the same methods used to analyze failure in collisions. However, hull-girder failure (i.e., "breaking the back" of a tanker) and hull tearing are specific to groundings and require special approaches. Hull-girder failures have been examined with the aid of increasingly powerful numerical models within the last five to six years. Issues studied include the contribution of dynamic effects to hull-girder collapse and the influence of friction between the hull and the seabed. Some scale-model and full-scale grounding tests have been conducted, and the results are in reasonable agreement with mathematical models.

Integration of a reliable fracture analysis approach into collision and grounding analyses would constitute a major advance in the analytical tools available to evaluate and design double-hull tankers. The approaches used today ensure that tankers have sufficient strength to withstand the loads encountered in regular operation, but there are no provisions for the loads encountered in accidents.[20] Similarly, outflow performance is based on tank subdivision only, and no consideration is given to the performance of the ship's structure in collisions and groundings. The development of tools that can be used to design tanker structures for good performance in accidents will be an important advance. Ongoing research has this objective, but much work is still required before research findings can be translated into practical design tools.

Other than work being conducted by the U.S. Navy,[21] the research of Wierzbicki and his coworkers at MIT is the main activity in the United States dedicated to the development of advanced analysis methods for ship structures (see Appendix L). Nearly all of Wierzbicki's effort is supported by industry, mainly from abroad. Most of the computer codes used to analyze nonlinear deformation of structures have been developed in this country, and the expertise needed to extend them to include the effects of collapse and fracture also resides in this country. The committee is concerned that important research opportunities may be missed because of the absence of any significant federal funding for this type of work.

---

[20]In-process accident behavior—such as that associated with collisions and groundings—is referred to as "crashworthiness" in the automotive and aviation industries.

[21]Commercial tanker designers and operators were not involved in either the planning or the execution of the U.S. Navy's $25 million project on the development of advanced double-hull tanker technology.

FINDINGS

Finding 1. The results of the comparative study commissioned by the committee indicate that—with the exception noted in Finding 2—double-hull tankers perform significantly better than single-hull tankers in preventing oil spills and mitigating oil outflow in the event of a collision or grounding.

- If a vessel experiences a collision or grounding that penetrates the outer hull, double-hull tankers are four to six times less likely than single-hull tankers to spill oil.
- Expected or average outflow is three to four times less with a double-hull compared to a single-hull tank vessel.
- The benefits of fitting double hulls to tank barges are at least as significant as they are for tankers.

Finding 2. The committee's analysis indicated that outflow performance is heavily influenced by the extent of subdivision within the cargo tanks of a double-hull tanker. Vessels with single-tank-across cargo tank arrangements (i.e., without longitudinal bulkheads) exhibit inferior performance with regard to both outflow and intact stability compared to other double-hull designs. Approximately half of the existing double-hull tankers of less than 160,000 DWT have single-tank-across configurations.

Double-hull tankers with single-tank-across cargo tank arrangements have approximately twice the average outflow of double-hull tankers with an oiltight centerline bulkhead in the cargo tanks.

Double-hull tankers with all single-tank-across cargo tanks perform more poorly than pre-MARPOL and MARPOL single-hull vessels with regard to extreme outflow.

All of the single-hull tankers analyzed in the committee's comparative study are inherently stable and do not experience intact stability problems. Four of the nineteen double-hull designs studied, however, may become unstable during loading and discharging operations. All four of these designs have single-tank-across cargo tank configurations.

Finding 3. There is no discernible difference between the survivability characteristics of single-hull and double-hull tankers, indicating that the MARPOL 73/78 subdivision requirements, supplemented by the raking bottom damage criterion included in the 1992 Amendments of MARPOL, ensure a high level of survivability.

Finding 4. Existing regulations governing the design of double-hull tankers do not ensure that the environmental advantages of double-hull tankers are consistently achieved. Despite their potentially poor performance, the aforementioned single-tank-across designs comply with existing regulations.

Finding 5. Representatives of the tanker industry generally believe that double-hull tankers can be operated safely, albeit with more resources and attention than needed for single-hull tankers, including increased attention to rigorous inspection and proper maintenance of interior coatings. Operational guidelines for double-hull tankers issued by the Tanker Structure Cooperative Forum in conjunction with the International Association of Classification Societies and the International Safety Guide for Oil Tankers and Terminals are considered adequate by most operators. These guidelines address issues such as access, inspection, maintenance, ventilation, and making ballast spaces inert.

Finding 6. Progress has been made in standardizing classification society regulations and in incorporating fatigue assessment and other analytical techniques into structural review of double-hull tanker design. However, some ship operators have expressed concern that minimum rule requirements do not result in a double-hull design that can be operated safely and economically throughout the expected life of a vessel.

Finding 7. Progress has been made in applying computational tools to simulate structural performance in accident scenarios. New data from grounding and collision tests have become available and can be used to calibrate analytical methods. Apart from research conducted by the U.S. Navy with limited commercial application, the U.S. government and U.S. industry have funded little research on ship structural responses to collisions and groundings.

REFERENCES

Chen, H.H., H.Y. Jan, J.F. Conlon, and D. Liu. 1993. New approach for the design and evaluation of double-hull tanker structures. SNAME Transactions 101:215–245.

Goodwin, M.J., J.C. Card, and J.S. Spencer. 1996. Study of double-hull tanker lolling and its prevention. Marine Technology 33(3):183–202.

Herbert Engineering Corporation. 1996. Comparative Study of Double-Hull vs. Single-Hull Tankers. Background paper prepared for the Marine Board Committee on Oil Pollution Act of 1990 (Section 4115) Implementation Review. San Francisco, Calif.: Herbert Engineering Corporation.

Huang, R.T. 1996. Microbial Influenced Corrosion in Cargo Tanks. Presented at NACE Corrosion 96 T-14B Marine Vessel Corrosion, Denver, Colorado, March 27–28.

International Maritime Organization (IMO). 1995. Interim Guidelines for the Approval of Alternative Methods of Design and Construction of Oil Tankers Under Regulation 13F(5) of Annex I of MARPOL 73/78. London: International Maritime Organization.

Marine Log. 1996. Are bugs eating some double-hulled tankers? p. 31, October.

Minorsky, V.U. 1959. An analysis of ship collisions with reference to protection of nuclear power plants. Journal of Ship Research 3(1):1–4.

Moore, C., J. Neumann, and D. Pippenger. 1996. Intact stability of double-hull tankers. Marine Technology 33(3):167–182.

Motoi, T., A. Murakami, A. Kohsaka, A. Kada, M. Unno, and T. Taniguchi. 1995. A new structural design of double-hull VLCC. Proceedings of MARIENV 95 1:191–197.

National Research Council (NRC). 1991. Tanker Spills: Prevention by Design. Marine Board Washington, D.C.: National Academy Press.

Tang-Jensen, P. 1995. Innovative structural design for high quality double-hull VLCCs. Proceedings of MARIENV 95,1:144–151.

Vredeveldt, A.W., and L.J. Wevers. 1992. Full-scale ship collision tests. Pp. 743–769 in Proceedings, First Conference on Marine Safety and Environment Ship Production, Delft, June 1–5, 1992. Delft: Delft University Press.

Vredeveldt, A.W., and L.J. Wevers. 1995. Full-scale grounding experiments. Pp. 11–112 in Proceedings of Conference on Predictions Methodology of Tanker Structural Failure and Consequential Oil Spill, Tokyo, April 1995. Tokyo: Association for Structural Improvements of the Shipbuilding Industry in Japan.

Wevers, L.J., J. van Vugt, and A.W. Vredeveldt. 1994. Full-scale six degrees of freedom motion measurements of two colliding 80 m long inland waterway tankers. Pp. 923–930 in Proceedings of the 10th International Conference on Experimental Mechanics, Lisbon, June 18–22, Rotterdam: A.A. Baldema.

7

Conclusions and Recommendations

The committee's conclusions and recommendations regarding the effects of Section 4115 of the Oil Pollution Act of 1990 (OPA 90) on pollution prevention, ship safety, and the composition and economic viability of the maritime oil transportation industry are presented in this chapter. The effects of the corresponding international rules and their interaction with Section 4115 are also addressed.

IMPLEMENTATION OF SECTION 4115

The passage of OPA 90 was a catalyst for a major change in the structural design of tank vessels; double-hull tankers are now the industry standard. Section 4115 and the MARPOL Regulations I3F and I3G (MARPOL 13F and 13G) differ in their approach and timing, although both sets of regulations require new tank vessels to have double hulls and provide for the gradual phaseout of existing single-hull vessels while the owners of oil tank vessels change or modify their capital plans to reflect new design requirements.

Section 4115 requires vessels operating in U.S. waters to have double hulls by 2015 at the latest; the international fleet governed by MARPOL is expected to be composed entirely of double-hull vessels (or approved alternatives) no later than 2023. Section 4115 restricts oil trade to the United States by vessels without double hulls according to a schedule based on vessel age but does not force such vessels into retirement. The first single-hull vessels reached mandatory phaseout in 1995. MARPOL 13G mandates the retirement of single-hull tankers from international trade at 30 years of age. Existing single-hull vessels are allowed to trade longer than under Section 4115 if they are of acceptable design. Thus, some single-hull vessels excluded from U.S. waters by OPA 90 may continue in trade to countries other than the United States until forced into retirement by MARPOL.

The U.S. Coast Guard (USCG) issued its final rules on operational and structural requirements to reduce the potential for oil pollution from existing single-hull vessels in July 1996 and January 1997, respectively. Vessels without double hulls operating in U.S. waters are not required to undertake any new structural measures before they are phased out. The operational measures took effect in November 1996. Given the lack of operational experience since the implementation of section 4115, the committee concluded that its impact merits reevaluation in approximately five years. This would permit assessment of the oil transportation industry's experience during the first five years of the mandatory phaseout of single-hull vessels under OPA 90. It would also coincide with the onset of a possible temporary vessel supply-demand imbalance. In the interim, efforts by the USCG to remedy deficiencies in its oil spill and port-state inspection databases would be beneficial for future assessments of the effect of OPA 90 (and other regulations and guidelines) on the protection of the marine environment and the quality of the tank vessel fleet operating in U.S. waters.

On the basis of the difficulties encountered in obtaining reliable oil spill data for the purposes of the present study, the committee identified some opportunities to enhance the USCG oil spill database. First, the USCG should recognize the importance of historical oil spill data as a primary indicator of achievements in the field of marine environmental protection. Adequate resource allocations for data gathering, data entry, and supervisory tasks would help ensure that reporting is complete and consistent throughout the USCG. Second, benefits could accrue if the public were given immediate access via the Internet to a simplified database[1] of all oil and chemical spills that have occurred since 1973 in locations that fall under USCG jurisdiction. If the availability of these data were widely publicized and the public encouraged to report discrepancies, possible errors could be scrutinized and corrections issued as necessary. Third, efforts to resolve discrepancies in historical records maintained by the USCG, the Minerals Management Service (MMS), the Oil Spill Intelligence Report, the International Tanker Owners Pollution Federation, and other entities would be beneficial in eliminating some of the apparent data anomalies of the type encountered by the committee. Agreement among different groups on consistent definitions of terms would be helpful in this regard.

---

[1]This simplified database might include information on the name and type of vessel or facility involved; vessel flag; date, time, and location of accident; age of vessel at time of accident; hull type (single, double, double sides, etc.); cause of accident; nature and type of commodity carried and volume spilled; and USCG case number.

Recommendation. The effects of the implementation of Section 4115 of OPA 90 should be reevaluated by an independent panel, possibly in about five years. The usefulness of such an evaluation will be greatly enhanced if the USCG initiates efforts now to ensure that adequate data will be available for future assessments.

Recommendation. The USCG should ensure that its oil spill database—including information on cause—is capable of facilitating the analysis of trends and the comparison of accidents involving oil spills. This would benefit the development of future regulations aimed at preventing oil spills and would facilitate industry planning.

Recommendation. The USCG should ensure that its port-state inspection database permits meaningful comparisons and analyses of current and future port-state activities, particularly in regard to identification and assessment of trends in the quality of the tank vessel fleet.

PROTECTION OF THE MARINE ENVIRONMENT

During the past five years, compared to earlier five-year periods, there has been a decline in the volume of oil spilled from vessel casualties in U.S. waters, together with an overall reduction in the number of oil spills of more than 100 gallons. The volumes released have been at historically low levels during the period 1991 to 1995. However, this decline in spills cannot be credited to Section 4115 because the mandatory phaseout of single-hull vessels commenced only in 1995, and the final rules on operational and structural measures for existing single-hull vessels had not been issued by the end of 1995.

In the view of the committee, a number of factors other than Section 4115 probably contributed to the recent reduction in oil spills. These include an increased awareness of the financial consequences of oil spills on the part of vessel owners and operators and a resulting increase in attention to policies and procedures aimed at eliminating vessel accidents; actions by port states to ensure the safety of vessels using their ports; increased efforts by classification societies to ensure that vessels under their classification meet or exceed existing requirements; improved audit and inspection programs by charterers; and the increased liability, financial responsibility, and other provisions of OPA 90. All of these actions are in process or emerging, as are vessel design requirements. It is therefore not possible to establish a direct correlation between any individual factors and the observed reduction in oil spillage.

On the basis of an analytical evaluation, the committee concluded that in the event of an accident involving a grounding or a collision, an effectively designed double-hull tanker will significantly reduce the expected outflow of oil compared to that from a single-hull tanker. Comparable analytical results were obtained for oceangoing barges. Inland and oceangoing barges together accounted for approximately half the total spillage and were involved in the majority of oil spills in U.S. waters between 1991 and 1995.

DESIGN OF DOUBLE-HULL TANK VESSELS

Since the passage of OPA 90, research on double-hull tanker design has provided significant insights into the impact of vessel design on double-hull tanker operations and pollution prevention capability, as measured in terms of expected oil outflow in the event of an accident. Overall, this research has demonstrated that effectively designed double-hull tankers and tank barges offer a significant improvement in environmental protection compared to that provided by single-hull vessels.

However, recent research—including the committee's probabilistic outflow analysis—has revealed possible intact stability and oil outflow problems with certain double-hull designs, all of which comply with existing design regulations of the International Maritime Organization (IMO) and major classification societies. More than half of the double-hull tankers less than 160,000 deadweight tons (DWT) that have entered service since the enactment of OPA 90 are of potentially problematic design because they have single-tank-across cargo tank arrangements. The tanker industry is entering a period of significant vessel retirements corresponding to the tanker building boom of the mid-1970s. The committee concluded that there is an urgent need to address double-hull design issues in time to impact the significant numbers of new double-hull vessels likely to enter service within the next few years and to prevent addition to the fleet of numerous vessels of unsatisfactory but approved design with a lifetime of 25 to 30 years.

Although intact stability problems can be avoided if adequate operational procedures are implemented, an unstable condition could still occur if the ship's crew fails to follow such procedures correctly. For this reason, a "design-only" solution has been suggested whereby the design of a double-hull vessel ensures stability at all times during cargo transfer operations. A possible disadvantage of the design-only approach is that it might limit possible options for designs with better environmental and damage stability performance (i.e., it may not be possible to optimize a design simultaneously for intact stability, damage stability, and environmental performance.) In addition, a design-only approach does not prevent the potentially unsafe operation of existing double-hull tankers susceptible to intact stability problems.

A possible approach to optimizing double-hull designs is to use performance-based criteria to evaluate environmental performance and other requirements. Such an approach is less prescriptive than conventional design rules. For example, rather than mandating centerline bulkheads in cargo tanks, performance-based criteria would give naval architects the opportunity to develop potentially superior designs and to address oil outflow performance and intact stability issues in a variety of ways. Thus, possible difficulties with centerline bulkheads in small tanker and barge designs (e.g., damage stability problems) could be avoided. A disadvantage of performance-based criteria in the present situation is that they are likely to take several years to develop, by which time more double-hull tankers with inferior environmental performance and intact stability problems may well have been built.

IMO has recently acted to address intact stability issues for double-hull tankers. An IMO circular provides guidance on operational measures needed to ensure adequate intact stability of existing vessels during load and discharge operations. MARPOL Draft Regulation I/25A(2) establishes a design-only requirement for intact stability for new vessels. Outflow regulations are currently under development at IMO.

Recommendation. The USCG should expand and expedite research efforts and cost-benefit evaluations necessary to develop rules appropriate for the design of double-hull tankers and tank barges. The following are of particular importance:

- Probabilistic analysis of oil outflow should be made an integral part of the design and review process for new double-hull tank vessels. Design requirements should ensure that all new double-hull tankers offer environmental performance at least equivalent to that provided by the IMO reference double-hull designs.[2]
- Design requirements should include an assessment of intact stability throughout the range of loading and ballasting conditions to identify potentially unstable conditions. Following the lead taken by IMO and to provide consistency with anticipated international requirements, adequate intact stability should be achieved by design.

Such rules should be implemented as soon as possible—if necessary in interim form—to ensure that all new double-hull tank vessels entering service do not pose a safety risk because of poor intact stability characteristics and have adequate internal subdivision to take full advantage of the spill-mitigating capabilities of double hulls.

Recommendation. The USCG should develop and implement operational procedures for existing double-hull tanker designs subject to intact stability problems. Such procedures should ensure adequate stability at all times during cargo transfer operations and should include appropriate crew training. Consistency between

[2]IMO has established a series of reference double-hull tanker designs for assessing environmental performance of alternative tanker arrangements. These designs were selected to exhibit a favorable oil outflow performance. Designs up to 150,000 DWT have an oiltight centerline bulkhead throughout the cargo block; the very large crude carrier (VLCC) reference ship is arranged with three-across cargo tanks.

procedures for vessels in U.S. waters and corresponding international procedures is highly desirable.

OPERATIONAL MAKEUP OF THE MARITIME OIL TRANSPORTATION INDUSTRY

The committee's analysis of the operational makeup of the maritime oil transportation industry indicated an increase in the proportion of double-hull tankers in the world fleet from 4 percent in 1990 to 10 percent in 1994, consistent with the requirements of OPA 90 and MARPOL 13F. Other changes in the world and U.S. trading fleets between 1990 and 1994—notably in trading patterns, age-related features, and vessel ownership—reflected both economic and regulatory factors and in some cases were a continuation of trends that predated OPA 90. These changes could not be definitively attributed to Section 4115 or to MARPOL 13F and 13G.

The committee concluded that the phaseout schedules of OPA 90 and MARPOL have not yet influenced the age of vessels calling on the United States. This is not unexpected because trading patterns observed before the implementation of OPA 90 indicate that few vessels over 25 years of age trade to the United States. However, some changes in the age distribution of the U.S. trading fleet are anticipated, particularly for the largest vessels. Factors contributing to such changes include the OPA 90 lightering zone and deepwater port exemption, the aging of the very large crude carrier (VLCC) fleet, and actions by other nations to discourage older vessels from calling on their ports.

Although the mandatory phaseout schedule of Section 4115 bans all single-hull tankers (without double bottoms or double sides) from U.S. trade after 2010, it is probable that under the deepwater port and lightering zone exemption, large single-hull vessels up to 30 years of age will operate to the United States through 2015. A large number of VLCCs constructed during the shipbuilding boom of the 1970s would have been excluded from U.S. waters between 1999 and 2003 under the normal OPA 90 phaseout schedule. However, the exemption allows them to trade until sometime between 2004 and 2008.[3] This situation is in contrast to that in 1994, when there were VLCCs more than 25 years of age in service.

The committee is concerned that in the future, there may be an overall deterioration in the quality of the VLCC fleet trading to the United States as the international VLCC fleet ages and other nations, such as Japan and Korea, introduce age restrictions on vessels calling on their ports. In the view of the committee, the United States needs to take appropriate measures to ensure that the older VLCCs operating under the OPA 90 deepwater port and lightering zone

[3]Vessels of pre-MARPOL design are required by Regulation 13G to use hydrostatically balanced loading (HBL) (or equivalents) to operate between 25 and 30 years of age.

exemption are adequately maintained and that their operation does not pose an unacceptable risk to the marine environment.

Recommendation. The U.S. Coast Guard should implement a vessel surveillance program to ensure that the physical condition, maintenance, and operating procedures of vessels permitted to discharge their cargo offshore, but barred from non-offshore ports by the phaseout provisions of Section 4115, are held to appropriate levels. For example, the frequency and standards of inspection defined in the Port State Inspection Program and applied to vessels using non-offshore ports might also be applied to vessels using lightering areas and the U.S. deepwater port.

ECONOMIC VIABILITY OF THE INTERNATIONAL TANKER INDUSTRY

The primary economic impact of the double-hull requirement on the international tanker fleet results from the higher capital and operating costs of double-hull compared to single-hull tankers. The cost to replace the current single-hull world trading fleet[4] with new double-hull tankers and operate them through a 20-year life cycle was estimated by the committee to be approximately $30 billion greater than building and operating an equivalent single-hull tanker fleet. Some shipowners are expected to take advantage of the lower capital cost of older single-hull tankers and adopt hydrostatically balanced loading (HBL) to extend the life of pre-MARPOL single-hull vessels beyond 25 years. However, such life extension requires expensive special surveys that will raise operating costs for older tankers, and the use of HBL will reduce cargo capacity and revenues.

Section 4115 will have little impact on the retirement of large single-hull tankers (150,000 DWT or more) that use HBL and are suitable for unloading within the lightering zones or at the deepwater port. Smaller single-hull tankers, particularly those for which unloading offshore is not economical, may be scrapped before the end of their economic life. Single-hull tankers of between 60,000 and 150,000 DWT (without double bottoms or double sides) will be excluded from trade to the United States when they reach 23 or 25 years of age, in accordance with the phaseout schedule of Section 4115.

Current shipyard capacity is ample to meet the world demand for new double-hull vessels and conversions. However, freight rate increases are anticipated as the industry transitions to a double-hull fleet. Given higher freight rates, it is expected that sufficient capital will be available to fund the conversion from a single-hull to a double-hull fleet. Therefore, the committee concluded that the international tanker industry is capable of transitioning to double-hull vessels in accordance with the requirements of OPA 90 and MARPOL.

---

[4]About 3,000 tankers aggregating 280 million DWT.

ECONOMIC VIABILITY OF THE JONES ACT TANK VESSEL FLEET

There are no alternatives to tank vessels for the movement of crude oil from Alaska to the lower 48 states. The Alaskan trade currently has sufficient tankers to meet projected demand until sometime between 2000 and 2006; the OPA 90 single-hull phaseout schedule may result in a requirement for double-hull vessels during this same period. If Alaskan North Slope production continues to decline as expected and no additional production is added, the need for new tankers will be short lived—probably less than 10 years.

The supply and demand situation in the coastal products trade is complex, but several logistical factors give rise to considerable uncertainty in future demand for vessels. There are some indications that domestic tank vessels in the coastal products trade are becoming less attractive economically than the alternatives—namely, pipelines and foreign tankers carrying imports of refined products. An increase in freight rates to induce replacement or conversion of single-hull vessels to double hulls might encourage pipeline expansion or an increase in product imports, thereby further reducing the demand for Jones Act vessels.

The impact of the double-hull requirement on the Jones Act tank vessel fleet is expected to be much greater than that on the international tanker fleet. Jones Act tank vessels are typically built with longer life expectancy than vessels in the international fleet and operate for 20 to 35 years (or more). Accordingly, they will generally reach their mandated retirement dates before the end of their economic life. In addition, the anticipated declines in demand in both the Alaskan crude oil trade and the coastal products trade may not provide for sufficient vessel life to recover investment in the double-hull vessels required by OPA 90, thereby discouraging new construction or conversions.

The committee concluded that there is an urgent need to address issues associated with domestic transportation capability—notably the impact on national defense and the ability to meet the energy needs of the Northeast under extraordinary circumstances such as severe winter weather and pipeline or refinery disruption. The effect of uncertainties over the future state of Jones Act market regulations should be included in the assessment.

Recommendation. The policy issues associated with the potential loss of domestic waterborne transportation capability should be carefully examined within the context of the double-hull mandate of Section 4115 and the committee's finding that properly designed double-hull vessels—including barges—are expected to offer enhanced environmental protection compared to single-hull designs. This examination should be undertaken by an independent body and should address the perspectives of all stakeholders, including tank vessel owners and operators, the oil industry and oil consumers, environmentalists, and state and federal regulators. The study should be initiated as soon as possible to ensure that policy determinations are made prior to potential disruptions or inefficient economic decisions.

APPENDICES

APPENDIX A

Biographies of Committee Members

Douglas C. Wolcott *(chair)* served as president of Chevron Shipping Company from 1984 until his retirement in 1994. During that time, Chevron had the largest oil company-owned fleet in the world, consisting of 40 oceangoing tankers with a total carrying capacity of 6 million deadweight tons, a smaller fleet of tugboats and barges, and 50 to 60 chartered vessels. Mr. Wolcott had been with Chevron Corporation (previously Standard Oil Company of California) since 1957, holding positions in oil-producing operations, the international fleet, traffic and chartering, and operations. Mr. Wolcott serves on the board of directors of the American Bureau of Shipping and of London and Overseas Freighters, Ltd. He has been chairman of the Oil Companies International Marine Forum, the American Institute of Merchant Shipping, and the Marine Preservation Association, and deputy chairman of the United Kingdom Protection and Indemnity Club. He holds a B.S. degree in engineering from the University of California at Berkeley and has completed graduate work in petroleum engineering at the University of Southern California.

Peter Bontadelli *(vice chair)* is administrator of the Office of Oil Spill Prevention and Response of the California Department of Fish and Game. He has primary authority for prevention, removal, abatement, response, containment, and cleanup efforts related to oil spills in the marine waters of California. His previous experience at the Department of Fish and Game included service as special assistant to the director, chief deputy director, and most recently, department director, a post he held for five years. During that time he served on various distinguished environmental panels, including the Pacific Flyway Council (where he was a former president), the North American Wetlands Conservation Council, the

Pacific Fishery Management Council, the Pacific States Marine Fisheries Commission, the International Association of Fish and Wildlife Agencies, and the Western Association of Fish and Wildlife Agencies. Mr. Bontadelli received his B.A. in political science from the University of California at Davis.

Lars Carlsson is president of Concordia Maritime AB, a Swedish shipping company that operates two ultralarge crude carriers and six very large crude carriers in cooperation with Stena Bulk AB. A senior executive in international shipping and trade since 1969, Mr. Carlsson is chair of the North Europe Committee of the American Bureau of Shipping, a council member of INTERTANKO (the International Association of Independent Tanker Owners), and a frequent participant in shipping conferences. He is an industry advocate for building and maintaining oil tankers to the highest standards and for providing these standards through voluntary quality classification. Mr. Carlsson holds a degree in business economy.

William R. Finger, is president of ProxPro, Inc., where he evaluates present and future prospects for the energy and oil industries. Prior to joining ProxPro in 1992, he served at the Exxon Company (USA) for 33 years, where he was responsible for evaluating the energy business environment and for representing Exxon in energy matters before the U.S. Congress and government agencies. He also represented the company in industry groups, including the National Petroleum Council, the Energy Modeling Forum, the American Petroleum Institute, and the Houston Economic Development Council. Mr. Finger received his B.S. degree from Virginia Polytechnic Institute and State University and is a registered professional engineer in the State of Louisiana.

Ran Hettena is president of the Maritime Overseas Corporation, the operating agent for the Overseas Shipholding Group, Inc. (OSG), and has been active in the shipping business for 40 years. At OSG he has been director and member of the Finance and Development Committee of the company and president of OSG Bulk Ships, Inc., a subsidiary that owns a U.S.-flag fleet. Mr. Hettena has served as trustee, treasurer, and chair of the Finance Committee of the Webb Institute of Naval Architecture; chair of the Tanker Subcommittee of the U.S. Department of Transportation Maritime Advisory Committee; member of the American Bureau of Shipping board managers; chair of the Committee of Gard in the Norwegian Protection and Indemnity Insurance Association; and director of the American Institute of Merchant Shipping. He has a B.S. degree from Columbia University and an M.S. in economics from New York University.

John W. Hutchinson, NAS/NAE, is the Gordon McKay Professor of Applied Mechanics at Harvard University, where he has been on the faculty since 1963. His research interests include solid mechanics, buckling of structures, and mechanical behavior and fracture of engineering materials. He has served as an editor for Academic Press and on the editorial boards of a number of journals. He is a member of the Defense Sciences Research Council of the Defense Advanced Research Projects Agency (U.S. Department of Defense) and a former member of the U.S. National Committee on Theoretical and Applied Mechanics. He is a member of the American Academy of Arts and Sciences, the Danish Center for Applied Mathematics and Mechanics, the American Society for Testing and Materials, and the American Ceramics Society. A former Guggenheim fellow, Dr. Hutchinson is a fellow of the American Society of Mechanical Engineers and the recipient of a number of professional awards. He has a B.S. in engineering mechanics from Lehigh University, a Ph.D. in mechanical engineering from Harvard University, and honorary doctoral degrees from the Swedish Royal Institute of Technology and the Technical University of Denmark.

Sally Ann Lentz is co-executive director and general counsel of Ocean Advocates, a nonprofit environmental organization dedicated to the protection of the marine environment. She represents environmental interests in national and international forums on ocean dumping, vessel source pollution, and other marine public policy issues and has served as adviser to U.S. delegations to the International Maritime Organization. Her previous positions have included staff attorney for Friends of the Earth and the Oceanic Society, as well as private practice. She holds a B.A. from Oberlin College, and a J.D. from the University of Maryland, and she has completed postgraduate study in European Community law. A member of the District of Columbia and Maryland bars, she served on the Committee on Tanker Vessel Design of the NRC Marine Board.

Donald Liu is senior vice president for technology at the American Bureau of Shipping (ABS), where he directs the international technology activities of the organization. In his 30-year career at ABS, Dr. Liu has held positions as senior vice president of the Technical Services Group, vice president of the Research and Development Division, assistant vice president, and chief research engineer. He has published and presented numerous technical papers on ships and marine loading and on computer analytical methods. Dr. Liu is a graduate of the U.S. Merchant Marine Academy (B.S.), the Massachusetts Institute of Technology (B.S. and M.S. degrees in naval architecture and marine engineering), and the University of Arizona (Ph.D. in mechanical engineering).

Dimitri A. Manthos has been president since 1962 of Admanthos Shipping Agency, Inc., of Stamford, Connecticut. Admanthos Shipping, founded in 1947, presently manages five modern product carriers in the U.S. trades and has a double-hull vessel under construction. Mr. Manthos has held senior positions with Tropic Drilling Company of Texas and other marine-oriented firms. He is a director of the U.K. Mutual Steamship Insurance Association and a member of the Det Norske Veritas North America Committee and the Bahamas Maritime Advisory

Council. He was a member and director of the Society of Maritime Arbitrators and served on the Ocean Industry Visiting Committee of the Massachusetts Institute of Technology (MIT), of which he is a life sustaining fellow. He holds a B.S. in naval architecture and marine engineering and an M.S. in shipping and shipbuilding management, both from MIT.

Henry Marcus, professor of marine systems at MIT, is chairman of the MIT Ocean Systems Management Program and the Naval Sea Systems Command (NAVSEA) Professor of Ship Acquisition. He holds a B.S. degree in naval architecture from the Webb Institute of Naval Architecture; M.S. degrees in naval architecture, shipbuilding, and shipping management from MIT; and a doctorate in business administration from Harvard University. Dr. Marcus chaired the Committee on Tank Vessel Design, which operated under the auspices of the National Research Council Marine Board and produced the 1991 report *Tanker Spills: Prevention by Design.*

Keith Michel is president of Herbert Engineering Corporation. In his 20 years with the company he has worked on design, specification development, and contract negotiations of container ships, bulk carriers, and tankers. Mr. Michel has served on industry advisory groups developing guidelines for alternative tanker designs, including groups advising the International Maritime Organization and the U.S. Coast Guard. His work has included development of methodology, vessel models, and oil outflow analysis. He was a project engineer for the U.S. Coast Guard report on oil outflow analysis for double-hull and hybrid tanker arrangements, which was part of the U.S. Department of Transportation's technical report on the Oil Pollution Act of 1990 (OPA 90) to Congress. He has also worked on the development of salvage software used by the U.S. and the Canadian Coast Guards, the U.S. Navy, the National Transportation Safety Board, the Maritime Administration, the American Bureau of Shipping, Lloyd's, and numerous oil and shipping companies. Mr. Michel holds a B.S. degree in naval architecture and marine engineering from the Webb Institute of Naval Architecture.

John H. Robinson is a consultant in marine science issues related to offshore oil development and transportation. Mr. Robinson retired from federal service after serving for 30 years in positions with the National Aeronautics and Space Administration (NASA) and the National Oceanic and Atmospheric Administration (NOAA). As director of the NOAA Gulf Program Office of the Office of the Chief Scientist, he directed NOAA research to assess the effects of marine oil spills and oilfield fires in the aftermath of the Persian Gulf war. Previously, as manager of the NOAA Hazardous Materials (HAZMAT) Response Division, he developed and managed the NOAA spill response and hazardous waste site research program, established regional scientific support programs in U.S. coastal areas, and served as scientific coordinator for the Ixtoc I oil drilling spill, the

Exxon Valdez, and other oil and chemical spills. While at HAZMAT, Mr. Robinson originated a program for computer-aided management of emergency operations. He received his B.S. in industrial engineering from Texas Technological University.

Ann Rothe is executive director of Trustees for Alaska, a nonprofit, public interest law firm representing environmental groups, Alaskan Native corporations, and others in the areas of natural resources and environmental protection. Prior to her current position, she was Alaska's regional representative to the National Wildlife Federation and assistant to the regional vice president of the National Audubon Society. After the *Exxon Valdez* grounding, she worked on state and federal legislation to improve oil spill prevention and response capabilities in Alaska and nationwide, and she was a principal organizer of the Regional Citizens Advisory Council for Prince William Sound. She has also served on the Research and Development Advisory Committee for the Marine Spill Response Corporation and the regional technical working group for outer continental shelf activities of the Minerals Management Service. Ms. Rothe has a B.S. in journalism and wildlife biology from Iowa State University.

David G. St. Amand is president and founder of Navigistics Consulting. An expert on shipping and petroleum economics, he has been a witness on shipping and petroleum economics, conducted extensive analyses of the Alaskan and foreign tanker trades, led a reengineering effort for the crude oil supply of a major oil company, and conducted studies on the regulatory and environmental effects of hydrocarbon vapor emission regulations. He was project manager for the development of vessel oil spill response plans for a number of shipowners and operators, and he has worked with owners, operators, and oil spill response contractors to ensure their compliance with OPA 90. He also serves on the Towing Safety Advisory Committee for the U.S. Coast Guard. Mr. St. Amand holds a B.S. in naval architecture and marine engineering from the Webb Institute of Naval Architecture and an M.B.A. from Dartmouth College.

Kirsi K. Tikka is associate professor at the Webb Institute. She was previously a senior analyst for tanker planning and economics at Chevron Shipping Company, where she performed economic analyses for marine transportation projects, including new vessel building projects, vessel charter evaluations, operation cost studies, transportation studies, and voyage economics. Dr. Tikka has degrees in mechanical engineering (M.S.) from the Helsinki University of Technology and in naval architecture and offshore engineering (M.S. and Ph.D.) from the University of California at Berkeley.

APPENDIX B

Oil Pollution Act of 1990 (P.L. 101-380), Section 4115

ESTABLISHMENT OF DOUBLE HULL REQUIREMENT FOR TANK VESSELS

(a) Double Hull Requirement.—Chapter 37 of title 46, United States Code, is amended by inserting after section 3703 the following new section:

"§ 3703a. Tank Vessel Construction Standards

"(a) Except as otherwise provided in this section, a vessel to which this chapter applies shall be equipped with a double hull—

"(1) if it is constructed or adapted to carry, or carries, oil in bulk as cargo or cargo residue; and

"(2) when operating on the waters subject to the jurisdiction of the United States, including the Exclusive Economic Zone.

"(b) This section does not apply to—

"(1) a vessel used only to respond to a discharge of oil or a hazardous substance;

"(2) a vessel of less than 5,000 gross tons equipped with a double containment system determined by the Secretary to be as effective as a double hull for the prevention of a discharge of oil; or

"(3) before January 1, 2015—

"(A) a vessel unloading oil in bulk at a deepwater port licensed under the Deepwater Port Act of 1974 (33 U.S.C. 1501 et seq.); or

"(B) a delivering vessel that is offloading in lightering activities—

"(i) within a lightering zone established under section 3715(b)(5) of this title; and

"(ii) more than 60 miles from the baseline from which the territorial sea of the United States is measured.

"(c) (1) In this subsection, the age of a vessel is determined from the later of the dates on which the vessel—

"(A) is delivered after original construction;

"(B) is delivered after completion of a major conversion; or

"(C) had its appraised salvage value determined by the Coast Guard and is qualified for documentation under section 4136 of the Revised Statutes of the United States (46 App. U.S.C. 14).

"(2) A vessel of less than 5,000 gross tons for which a building contract or contract for major conversion was placed before June 30, 1990, and that is delivered under that contract before January 1, 1994, and a vessel of less than 5, 000 gross tons that had its appraised salvage value determined by the Coast Guard before June 30, 1990 and that qualifies for documentation under section 4136 of the Revised Statutes of the United States (46 App. U.S.C. 14) before January 1, 1994, may not operate in the navigable waters or the Exclusive Economic Zone of the United States after January 1, 2015, unless the vessel is equipped with a double hull or with a double containment system determined by the Secretary to be as effective as a double hull for the prevention of a discharge of oil.

(3) A vessel for which a building contract or contract for major conversion was placed before June 30, 1990, and that is delivered under that contract before January 1, 1994, and a vessel that had its appraised salvage value determined by the Coast Guard before June 30, 1990, and that qualifies for documentation under section 4136 of the Revised Statutes of the United States (46 App. U.S.C. 14) before January 1, 1994, may not operate in the navigable water or Exclusive Economic Zone of the United States unless equipped with a double hull—

"(A) in the case of a vessel of at least 5,000 gross tons but less than 15,000 gross tons—

"(i) after January 1, 1995, if the vessel is 40 years old or older and has a single hull, or is 45 years old or older and has a double bottom or double sides;

"(ii) after January 1, 1996, if the vessel is 39 years old or older and has a single hull, or is 44 years old or older and has a double bottom or double sides;

"(iii) after January 1, 1997, if the vessel is 38 years old or older and has a single hull, or is 43 years old or older and has a double bottom or double sides;

"(iv) after January 1, 1998, if the vessel is 37 years old or older and has a single hull, or is 42 years old or older and has a double bottom or double sides;

"(v) after January 1, 1999, if the vessel is 36 years old or older and has a single hull, or is 41 years old or older and has a double bottom or double sides;

"(vi) after January 1, 2000, if the vessel is 35 years old or older and has a single hull, or is 40 years old or older and has a double bottom or double sides; and

"(vii) after January 1, 2005, if the vessel is 25 years old or older and has a single hull, or is 30 years old or older and has a double bottom or double sides;

"(B) in the case of a vessel of at least 15,000 gross tons but less than 30,000 gross tons—

"(i) after January 1, 1995, if the vessel is 40 years old or older and has a single hull, or is 45 years old or older and has a double bottom or double sides;

"(ii) after January 1, 1996, if the vessel is 38 years old or older and has a single hull, or is 43 years old or older and has a double bottom or double sides;

"(iii) after January 1, 1997, if the vessel is 36 years old or older and has a single hull, or is 41 years old or older and has a double bottom or double sides;

"(iv) after January 1, 1998, if the vessel is 34 years old or older and has a single hull, or is 39 years old or older and has a double bottom or double sides;

"(v) after January 1, 1999, if the vessel is 32 years old or older and has a single hull, or is 37 years old or older and has a double bottom or double sides;

"(vi) after January 1, 2000, if the vessel is 30 years old or older and has a single hull, or is 35 years old or older and has a double bottom or double sides;

"(vii) after January 1, 2001, if the vessel is 29 years old or older and has a single hull, or is 34 years old or older and has a double bottom or double sides;

"(viii) after January 1, 2002, if the vessel is 28 years old or older and has a single hull, or is 33 years old or older and has a double bottom or double sides;

"(ix) after January 1, 2003, if the vessel is 27 years old or older and has a single hull, or is 32 years old or older and has a double bottom or double sides;

"(x) after January 1, 2004, if the vessel is 26 years old or older and has a single hull, or is 31 years old or older and has a double bottom or double sides; and

"(xi) after January 1, 2005, if the vessel is 25 years old or older and has a single hull, or is 30 years old or older and has a double bottom or double sides; and

"(C) in the case of a vessel of at least 30,000 gross tons—
"(i) after January 1, 1995, if the vessel is 28 years old or older and has a single hull, or is 33 years old or older and has a double bottom or double sides;
"(ii) after January 1, 1996, if the vessel is 27 years old or older and has a single hull, or is 32 years old or older and has a double bottom or double sides;
"(iii) after January 1, 1997, if the vessel is 26 years old or older and has a single hull, or is 31 years old or older and has a double bottom or double sides;
"(iv) after January 1, 1998, if the vessel is 25 years old or older and has a single hull, or is 30 years old or older and has a double bottom or double sides;
"(v) after January 1, 1999, if the vessel is 24 years old or older and has a single hull, or is 29 years old or older and has a double bottom or double sides; and
"(vi) after January 1, 2000, if the vessel is 23 years old or older and has a single hull, or is 28 years old or older and has a double bottom or double sides.
"(4) Except as provided in subsection (b) of this section—
"(A) a vessel that has a single hull may not operate after January 1, 2010; and
"(B) a vessel that has a double bottom or double sides may not operate after January 1, 2015."

(b) Rulemaking.—The Secretary shall, within 12 months after the date of the enactment of this Act, complete a rulemaking proceeding and issue a final rule to require that tank vessels over 5,000 gross tons affected by section 3703a of title 46, United States Code, as added by this section, comply until January 1, 2015, with structural and operational requirements that the Secretary determines will provide as substantial protection to the environment as is economically and technologically feasible.

(c) Clerical Amendment.—The analysis for chapter 37 of title 46, United States Code, is amended by inserting after the item relating to section 3703 the following:

"3703a. Tank Vessel Construction Standards."

(d) Lightering Requirements.—Section 3715(a) of title 46, United States Code, is amended—
(1) in paragraph (1), by striking "; and" and inserting a semicolon;
(2) in paragraph (2), by striking the period and inserting "; and"; and
(3) by adding at the end the following:
"(3) the delivering and the receiving vessel had on board at the time

of transfer, a certificate of financial responsibility as would have been required under section 1016 of the Oil Pollution Act of 1990, had the transfer taken place in a place subject to the jurisdiction of the United States;

"(4) the delivering and the receiving vessel had on board at the time of transfer, evidence that each vessel is operating in compliance with section 311(j) of the Federal Water Pollution Control Act (33 U.S.C. 1321(j)); and

"(5) the delivering and the receiving vessel are operating in compliance with section 3703a of this title."

(e) Secretarial Studies.—

(1) Other Requirements.—Not later than 6 months after the date of enactment of this Act, the Secretary shall determine, based on recommendations from the National Academy of Sciences or other qualified organizations, whether other structural and operational tank vessel requirements will provide protection to the marine environment equal to or greater than that provided by double hulls, and shall report to the Congress that determination and recommendations for legislative action.

(2) Review and Assessment.—The Secretary shall—

(A) Periodically review recommendations from the National Academy of Sciences and other qualified organizations on methods for further increasing the environmental and operational safety of tank vessels;

(B) not later than 5 years after the date of enactment of this Act, assess the impact of this section on the safety of the marine environment and the economic viability and operational makeup of the maritime oil transportation industry; and

(C) report the results of the review and assessment to the Congress with recommendations for legislative or other action.

(f) Vessel Financing.—Section 1104 of the Merchant Marine Act of 1936 (46 App. U.S. C. 1274) is amended—

(1) by striking "Sec. 1104." and inserting "Sec. 1104A."; and

(2) by inserting after section 1104A (as redesignated by paragraph (1)) the following:

"Sec. 1104B. (a) Notwithstanding the provision of this title, except as provided in subsection (d) of this section, the Secretary, upon the terms the Secretary may prescribe, may guarantee or make a commitment to guarantee, payment of the principal of and interest on an obligation which aids in financing and refinancing, including reimbursement to an obligor for expenditures previously made, of a contract for construction or reconstruction of vessel or vessels owned by citizens of the United States which are designed and to be employed for commercial use in the coastwise or intercoastal trade or in foreign trade as defined in section 905 of this Act if—

"(1) the construction or reconstruction by an applicant is made necessary to replace vessels the continued operation of which is denied by virtue of the imposition of a statutorily mandated change in standards for the operation of vessels, and where, as a matter of law, the applicant would otherwise be denied the right to continue operating vessels in the trades in which the applicant operated prior to the taking effect of the statutory or regulatory change;

"(2) the applicant is presently engaged in transporting cargoes in vessels of the type and class that will be constructed or reconstructed under this section, and agrees to employ vessels constructed or reconstructed under this section as replacements only for vessels made obsolete by changes in operating standards imposed by statute;

"(3) the capacity of the vessels to be constructed or reconstructed under this title will not increase the cargo carrying capacity of the vessels being replaced;

"(4) the Secretary has not made a determination that the market demand for the vessel over its useful life will diminish so as to make the granting of the guarantee fiduciarily imprudent; and

"(5) the Secretary has considered the provisions of section 1104A (d)(1)(A) (iii), (iv), and (v) of this title.

"(b) For the purposes of this section—

"(1) the maximum term for obligations guaranteed under this program may not exceed 25 years;

"(2) obligations guaranteed may not exceed 75 percent of the actual cost or depreciated actual cost to the applicant for the construction or reconstruction of the vessel; and

"(3) reconstruction cost obligations may not be guaranteed unless the vessel after reconstruction will have a useful life of at least 15 years.

"(c) (1) The Secretary shall by rule require that the applicant provide adequate security against default. The Secretary may, in addition to any fees assessed under section 1104A(e), establish a Vessel Replacement Guarantee Fund into which shall be paid by obligors under this section—

"(A) annual fees which may be an additional amount on the loan guarantee fee in section 1104A(e) not to exceed an additional 1 percent; or

"(B) fees based on the amount of the obligation versus the percentage of the obligor's fleet being replaced by vessels constructed or reconstructed under this section.

"(2) The Vessel Replacement guarantee Fund shall be a subaccount in the Federal Ship Financing Fund, and shall—

"(A) be the depository for all moneys received by the Secretary under sections 1101 through 1107 of this title with respect to guarantee or commitments to guarantee made under this section;

"(B) not include investigation fees payable under section 1104A(f) which shall be paid to the Federal Ship Financing Fund; and

"(C) be the depository, whenever there shall be outstanding any notes or obligations issued by the Secretary under section 1105(d) with respect to the Vessel Replacement Guarantee Fund, for all moneys received by the Secretary under sections 1101 through 1107 from applicants under this section.

"(d) The program created by this section shall, in addition to the requirements of this section, be subject to the provisions of sections 1101 through 1103; 1104A(b) (1), (4), (5), (6); 1104A(e); 1104A(f); Financing Fund is not liable for any guarantees or commitments to guarantee issued under this section."

APPENDIX C

Questionnaires

QUESTIONNAIRE FOR OWNERS AND OPERATORS OF DOUBLE-HULL TANK VESSELS

I. Operation of double-hull tankers
   1. What is your experience with operational safety of double-hull tankers in regard to:
      - stability during loading and discharging
      - safe access to ballast spaces
      - ventilation of ballast spaces
      - any other safety issues that need to be addressed
   2. Are there significant differences in cargo operations between double-hull and single-hull tankers?
   3. Have you established operational procedures specifically for double-hull tankers?

II. Inspection and maintenance of double-hull tankers
   1. Please provide information on structural and tank coating inspection frequencies and practices on double-hull tankers.
   2. What is your experience with different types of coating in ballast spaces? Have you encountered significant corrosion problems? If so, please describe.
   3. What are your current practices with regard to ballast tank coatings (include type, number of coats, thicknesses)? From your experience, what is the expected life of the coatings?
   4. Do any of your maintenance and inspection practices for single-hull tankers differ from those used on double-hull tankers?

III. Design of double-hull tankers
   1. Have you had any structural problems on double-hull tankers? Please provide information on the type of problems.
   2. What is your experience with high-strength steel construction?
   3. What design changes would you suggest in future double-hull tankers?

IV. Fleet information
   1. Please provide the number and size characteristics of double-hull tankers in your fleet.
   2. Please note if any of your operation experiences are specific to certain sizes of double-hull tankers.

V. General
   1. Based on your experience, what are the advantages and disadvantages of double-hull tankers as compared to single-hull tankers?

QUESTIONNAIRE FOR SHIPYARD OPERATORS, CLASSIFICATION SOCIETIES, AND MARINE ARCHITECTS

I. Design characteristics
   1. See the attached ship characteristics form for double-hull tankers. Kindly complete the form for double-hull tankers that have been built, or are under construction, or on order in your yard. An example of a completed form is provided for your guidance in completing the form.
   2. Additionally, what is the percentage of high-strength steel used in each design?
   3. What design changes do you foresee in future double-hull tankers?

II. Producibility
   1. In comparing the producibility of single-hull and double-hull designs of 90,000, 150,000, and 280,000 DWT sizes, please provide an estimate of the differences (in absolute terms or on a percentage basis) of labor hours or cost between single-hull and double-hull construction for:
      a) steel fabrication
      b) machinery/outfitting
      c) coatings (include type and extent)
      d) total construction time (keel laying to delivery)
      e) any other comparative data related to construction or producibility
   2. Please describe any particular problems in double-hull construction versus single-hull construction.

III. Maintenance
   1. Accessibility of spaces: what has been your experience relative to ease of access of spaces in double-hull tankers versus single-hull tankers?
   2. Ability to gas-free spaces: what has been your experience in the ability to gas-free spaces in double-hull designs for the safe entry of personnel?
   3. Maintenance-related problems: please describe any maintenance-related problems experienced with double-hull tankers.

OWNERS AND OPERATORS OF DOUBLE-HULL VESSELS WHO RECEIVED QUESTIONNAIRES

Name	*Location*
Acomarit (UK), Ltd.	Glasgow, United Kingdom
Acomarit Service S.A.	Geneva, Switzerland
A.P. Moller Company	Copenhagen, Denmark
Bergesen D.Y. A/S	Oslo, Norway
Bona Shipping A/S	Oslo, Norway
Ceres Hellenic Shipping Enterprises, Ltd.	Piraeus, Greece
Chevron Shipping Co.	San Francisco, California
Conoco Shipping Co.	Houston, Texas
Eletson Corporation	Piraeus, Greece
Essar Shipping, Ltd.	Madras, India
Frontline AB	Stockholm, Sweden
Gotass-Larsen, Ltd.	London, United Kingdom
Knutsen O.A.S. Shipping A.S.	Haugesund, Norway
Mitsui O.S.K. Lines	Tokyo, Japan
Mobil Shipping and Transportation Co.	Fairfax, Virginia
Mowinckels Rederi A/S	Bergen, Norway
Naess Shipping	Amsterdam, the Netherlands
Neptune Orient Lines, Ltd.	Singapore
Neste Oy	Esbo, Finland
Ocean Technologies, Ltd.	Ft. Lee, New Jersey
Teekay Shipping (Canada), Ltd.	Vancouver, Canada
Torre Britanica	Caracas, Venezuela
Tschundi & Eitzen	Lysaker, Norway
Ugland Tanker A/S	Grimstad, Norway

SHIPYARDS AND VESSEL DESIGNERS WHO RECEIVED QUESTIONNAIRES

Name	*Location*
Astilleros Espanoles (Puerto Real)	Cadiz, Spain
Avondale Industries	New Orleans, Louisiana
Bremer Vulkan Werft	Bremen, Germany
Chantier de l'Atlantique	Paris, France
China Shipbuilding Corporation	Taipei, Taiwan
Daewoo Heavy Industries, Ltd.	Kyungnam, South Korea
Daewoo Heavy Industries, Ltd.	Seoul, South Korea
Fincantieri	Trieste, Italy
Halla Engineering and Heavy Industries, Ltd.	Seoul, South Korea
Hanjin Heavy Industries Co., Ltd.	Pusan, South Korea

SHIPYARDS AND VESSELS OWNERS—*continued*

Name	*Location*
Hitachi Zosen Corporation	Osaka City, Japan
Hyundai Heavy Industries Co., Ltd.	Ulsan, Korea
Imabari Shipbuilding Co., Ltd.	Kagawa, Japan
Ishikawajima-Harima Heavy Industries, Ltd.	Tokyo, Japan
Kawasaki Heavy Industries, Ltd.	Hyogo, Japan
Kvaerner Warnow Werft	Warnermunde, Germany
Mitsubishi Heavy Industries Co., Ltd.	Yokohama, Japan
Mitsui Engineering and Shipbuilding Co., Ltd.	Chiba, Japan
Namua Shipbuilding Co., Ltd.	Imari City, Japan
Newport News Shipbuilding	Newport News, Virginia
NKK Corporation	Yokohama, Japan
Odense Steel Shipyard	Odense, Denmark
Onomichi Dockyard Co., Ltd.	Hiroshima, Japan
Oshima Shipbuilding Company, Ltd.	Nagasaki, Japan
Samsung Heavy Industries	Kyungnam, South Korea
Sanoyas Hishino Meisho Corporation	Okayama, Japan
Sasebo Heavy Industries Co., Ltd.	Nagasaki, Japan
Shin Kurushima Dockyard Co., Ltd.	Ochi-gin, Japan
Sumitomo Heavy Industries Co., Ltd.	Kanagawa, Japan
Tsuneishi Shipbuilding Co., Ltd.	Hiroshima, Japan

CLASSIFICATION SOCIETIES THAT RECEIVED QUESTIONNAIRES

Name	*Location*
American Bureau of Shipping	New York, New York
Bureau Veritas	Paris, France
Det Norske Veritas	Hovik, Norway
Germanischer Lloyd	Hamburg, Germany
Korean Register of Shipping	Taejon, South Korea
Lloyd's Register of Shipping	London, United Kingdom
Nippon Kaiji Kyokai	Tokyo, Japan
Polish Register of Shipping	Gdansk, Poland
Registro Italiano Navale	Genoa, Italy
Maritime Register of Shipping (formerly Russian Register)	St. Petersburg, Russia

MARINE ARCHITECTS WHO RECEIVED QUESTIONNAIRES

Name	*Location*
Beresford House, Town Quay	Southampton, United Kingdom
George G. Sharp, Inc.	New York, New York
John J. McMullen Associates, Inc.	New York, New York
M. Rosenblatt & Son, Inc.	New York, New York
Three Quays Marine Services	London, United Kingdom

APPENDIX

D

Committee Meetings and Activities

COMMITTEE MEETING

First Committee Meeting, March 9–10, 1995, Washington, D.C.

The following presentations were given by guest speakers:

Overview of Coast Guard objectives and status of Oil Pollution Act of 1990 (P.L. 101-380) (OPA 90) and International Convention for the Prevention of Pollution from Ships, adopted in 1973 and amended in 1978 (MARPOL) implementation
Capt. Dennis Bryant, U.S. Coast Guard

Second Committee Meeting, June 12–13, 1995, Irvine, California

The following presentations were given by guest speakers:

Effect of OPA 90, Section 4115 on tank vessel inspection—procedures and concerns
John Ferguson, Deputy Chief Surveyor, Lloyd's Register of Shipping
Linwood Poindexter, Vice President, North American Region, ABS America

Effect of OPA 90, Section 4115 on insurance of tank vessels operating in U.S. waters
John Hickey, President, American Hull Insurance Syndicate

Effect of OPA 90, Section 4115 on tank vessel sale, purchase, and chartering patterns
Samuel Jones, President, Mallory Jones Lynch Flynn and Associates
John Loucas, Vice President, McQuilling Brokerage Partners

Effect of OPA 90, Section 4115 on tank vessel design, producibility, and cost
Yoshiaki Sezaki, Manager, Design Division, Shipbuilding Headquarters, Hitachi Zosen Corporation

Third Committee Meeting, October 9–11, 1995, Washington, D.C.

The following presentations were given by guest speakers:

Industry perspectives on OPA 90, Section 4115
Miles Kulukindis, Chairman, INTERTANKO[1] *and President and Chief Executive Officer, London and Overseas Freighters Ltd.*
Dagfinn Lunde, Deputy Managing Director, INTERTANKO
Eric Shawyer, Chairman and Chief Executive, E.A. Gibson Shipbrokers Ltd.
Martin Stopford, Managing Director, Clarkson Research Studies

Insurance perspectives on OPA 90, Section 4115
Nickolai Herlofoson, Managing Director, Gard P&I

Shipbuilding perspectives on OPA 90, Section 4115
Jung Nam Lee, Executive Vice President, Sun Jong Park, and Byung O. Kim, Hyundai Heavy Industries Co., Ltd.
Richard Neilson, Manager, Research and Concept Design, Newport News Shipbuilding

Financing perspectives on OPA 90, Section 4115
James Grubbs, Senior Industry Analyst, Global Shipping Group, Citibank, N.A.
John Newbold, Division Executive, Global Shipping Division

Coast Guard comments on implementation of OPA 90
RADM James Card, U.S. Coast Guard

Fourth Committee Meeting, February 1–2, 1996, Irvine, California

The following presentations were given by guest speakers:

Engineering, maintenance, and inspection aspects of double-hull fleet operations
Richard Whiteside, Engineering Manager, British Petroleum

Double-hull tanker operations
Mitch Koslow, General Manager, Keystone Shipping

[1]The International Association of Independent Tanker Owners.

Measures to reduce oil spills from existing tank vessels without double hulls—highlights and rationale for U.S. Coast Guard proposed rule
Paul Cojeen, U.S. Coast Guard

Determining change in quality of the tanker fleet
Jack Klingel, U.S. Coast Guard

Fifth Committee Meeting, April 25–26, 1996, Washington, D.C.

The following presentations were given by guest speakers:

Barge industry perspectives on OPA 90, Section 4115
Tom Allegretti, President, and Jennifer Kelly, Vice President, American Waterways Operators
Jon Wales, Vice President, Reinauer Transportation

Sixth Committee Meeting, June 17–18, 1996, Woods Hole, Massachusetts

STUDIES PERFORMED UNDER SUBCONTRACT

Comparative Study of Double-Hull vs. Single-Hull Tankers. 1996. Herbert Engineering Corporation, San Francisco.
Forecast of Petroleum Flows and Tanker Needs, 1995–2005. 1995. PIRA Energy Group, New York.
Seasonality Trends. 1996. William Gassman. Department of Ocean Engineering, Massachusetts Institute of Technology, Cambridge.
Tanker Supply and Demand Analysis, Task 1 and Task 2. 1996. Navigistics Consulting, Boxborough, Mass.
Tankships Calling on U.S. Ports, 1990 and 1994. 1995. Institute for Shipping Analysis, Göteborg, Sweden.
U.S. Coast Guard Oil Spill Data, 1978–1995. 1996. Robert Brulle. Fairfax, Va.

OTHER ACTIVITIES

Testimony before U.S. Senate Committee on Environment and Public Works. Douglas C. Wolcott, June 4, 1996.

APPENDIX

E

Supplementary Data on Vessel Ownership for the U.S. Trading Fleet

Table E-1 shows the change in age from 1990 to 1994 on the basis of ownership category for vessels in the U.S. trading fleet. For government-owned vessels there was an increase in age of 0.38 year, but this was more than offset by the decrease of 1.02 years in the age of the numerous independent fleet and 0.99 year in the small oil company fleet. When figures based on the number of port calls are considered, rather than just the fleet itself, as shown in Table E-2, the change is more pronounced: 1.66 years average reduction, with a 0.35-year decrease for governments, a 1.49-year decrease for independents, and a 2.51-year decrease for oil companies. The new vessels are clearly making more calls on average than the older ones.

TABLE E-1 Changes in Age of the U.S. Trading Fleet Based on Individual Ships by Ownership Category, 1990–1994

	Government			Independents			Oil Companies		
Age Range	1990	1994	Change	1990	1994	Change	1990	1994	Change
0–4	1.83	1.78	–0.05	2.23	2.07	–0.16	2.46	1.75	–0.71
5–9	7.53	7.53	0.00	7.48	6.75	–0.73	8.01	6.46	–1.55
10–14	12.59	11.65	–0.94	12.58	12.41	–0.17	12.16	12.41	+0.25
15–19	15.00	17.18	+2.18	16.00	17.60	+1.60	16.06	17.63	+1.57
20–24				20.48	20.74	+0.26	20.87	20.52	–0.35
25 or more				35.97	33.44	–2.53	30.45	37.67	+7.22
Average	9.05	9.43	+0.38	11.55	10.53	–1.02	13.50	12.51	–0.99

TABLE E-2 Changes in Age of the U.S. Trading Fleet Based on Port Calls by Ownership Category, 1990–1994

	Government			Independents			Oil Companies		
Age Range	1990	1994	Change	1990	1994	Change	1990	1994	Change
0–4	2.13	2.39	+0.28	2.22	2.04	–0.18	2.46	1.65	–0.81
5–9	7.98	7.51	–0.47	7.64	6.79	–0.85	8.01	6.53	–1.48
10–14	12.29	11.79	–0.50	12.43	12.53	+0.10	12.16	12.34	+0.18
15–19	15.00	17.26	+2.26	15.90	17.49	+1.59	16.06	17.58	+1.52
20–24				20.87	20.97	+0.10	20.87	20.41	–0.46
25 or more				36.24	34.16	–2.08	30.45	37.14	+6.69
Total	9.10	8.85	–0.35	11.64	10.15	–1.49	12.92	10.41	–2.51

In looking beyond averages at size categories, it can be seen that the average age of government vessels has been decreasing for smaller vessels (≤150,000 DWT [deadweight tons]) and becoming more diverse for larger vessels (>150,000 DWT). The same broadly applies to oil companies. Independents show a split tendency, with old and young ships predominating in large sizes but few in the middle age bands, and a broad balance in the smaller sizes (see Table E-3).

TABLE E-3 Size in Million DWT of the U.S. Trading Fleet by Age Range and Ownership Category, 1990 and 1994

	Government				Independents				Oil Companies			
	0–150,000 DWT		> 150,000 DWT		0–150,000 DWT		> 150,000 DWT		0–150,000 DWT		> 150,000 DWT	
Age Range	1990	1994	1990	1994	1990	1994	1990	1994	1990	1994	1990	1994
0–4	0.34	1.98	0	2.61	9.93	27.91	3.93	7.90	0.89	3.42	0.15	0.28
5–9	1.25	1.58	0	0.61	12.91	13.55	2.99	1.62	2.05	1.19	0.26	0
10–14	0.65	0.95	1.71	0.59	17.12	14.74	17.12	1.45	2.60	1.32	7.28	0
15–19	0.22	1.30	0	2.10	17.96	17.56	18.72	21.79	3.51	1.87	10.99	3.66
20–24					2.04	4.70	0	5.23	0.16	1.62	0.31	3.50
25 or more					0.69	0.92	0.90	0	0.20	0.14	0.27	0
Total	2.46	5.81	1.71	5.91	60.65	79.38	43.64	37.99	9.41	9.55	19.25	7.45

APPENDIX
F

Methodology for Determining the International Tanker Supply

The international tanker fleet is comprised of a wide mix of vessels varying in size, type, and age. Vessel sizes range from less than 10,000 deadweight tons (DWT) up to 550,000 DWT. Vessel types are differentiated by hull type (e.g., single, double) and ballast tank configuration (e.g., segregated, protectively located). Tankers are designed for specific commercial purposes, such as crude oil trades only, product trades, and the shipment of other commodities including combinations of oil and ore; oil, bulk cargoes, and ore; chemicals; asphalt; acid; edible oils and juices; and liquefied gases. Because of this diversity of vessel types and purposes, it is important to list the assumptions used to arrive at the international tanker supply (i.e., number of tankers and total tonnage).

After discussions with invited experts (including Dr. Stopford of Clarkson Research Studies Ltd., Mr. Shawyer of E.A. Gibson Shipbrokers Ltd., Mr. Lunde of International Association of Independent Tanker Owners (INTERTANKO), and Mr. Kulukundis of INTERTANKO and London and Overseas Freighters), the committee determined that the following assumptions formed a reasonable basis for its analysis of the international fleet:

- Clarkson's existing and newbuilding tanker databases are accurate after adjustments based on input from Lloyd's, the American Bureau of Shipping, and Det Norske Veritas. (These adjustments were provided to Clarkson for incorporation into subsequent versions of its databases.)
- Only tankers of more than 10,000 DWT are included.
- The International Convention for the Prevention of Pollution from Ships, adopted in 1973 and amended in 1978, Regulations I/13F and I/13G (MARPOL 13F and 13G) are adopted by all countries except the United

States. Therefore, mandatory vessel retirements are determined by MARPOL and not by the Oil Pollution Act of 1990 (P.L. 101-380) (OPA 90), which bans non-double-hull tankers from U.S. waters no later than 2015 but does not preclude them from trading with other nations.
- Because the committee is examining the impact of mandated retirements only, economic or other types of retirements not driven by MARPOL 13G are excluded.
- MARPOL retires tankers based on delivery date. The Clarkson database used by the committee gives the year, but not the month, of delivery. Therefore, in the year of mandatory retirement, half of the vessel's deadweight is deleted from the supply. The full deadweight is deleted from all subsequent years.
- Tankers without double hulls that comply with MARPOL are retired at 30 years of age. (Double-hull tankers are unaffected by MARPOL retirement provisions.) MARPOL tankers are identified as those built from 1983 onward, plus those identified in Clarkson's database as having double sides or double bottoms, or having been built after 1979 and having segregated ballast tanks.
- All pre-MARPOL tankers will use hydrostatically balanced loading (HBL) in order to trade to 30 years of age. HBL results in an 8 percent loss in cargo carrying capacity for all pre-MARPOL tankers from age 25 through 30.
- Because of the OPA 90 lightering exemption, no tankers of more than 120,000 DWT are excluded from trade to the United States until 2015.
- Because of the current tight market for chemical tankers, these vessels are excluded from the supply of product tankers.
- Oil-bulk-ore carriers are identified separately.
- All government-owned military supply tankers, specialty tankers, and gas tankers are excluded from the supply.
- The deadweight of newbuildings is added to the supply based on the month and year of delivery. The full deadweight is included in all subsequent years. When delivery month is not specified, delivery is assumed to be at the beginning of the fourth quarter (i.e., 25 percent of the vessel's capacity is added to the supply in the year of delivery).

The committee's supply analysis is based on tanker fleet statistics (including orders) as of October 1, 1995.

APPENDIX G

Freight Rate Mechanism in the Short Run: A Theoretical Approach

Tanker freight rates are highly volatile. The rates paid for very large crude carriers (VLCCs) over a 35-year period, for example, have shown variations of 1,000 percent (Figure G-1). Aside from irregular shocks resulting from unexpected weather changes and political developments, freight rate volatility can be attributed to two principal causes: (1) the character of tanker supply, and (2) seasonal variations in tanker demand.

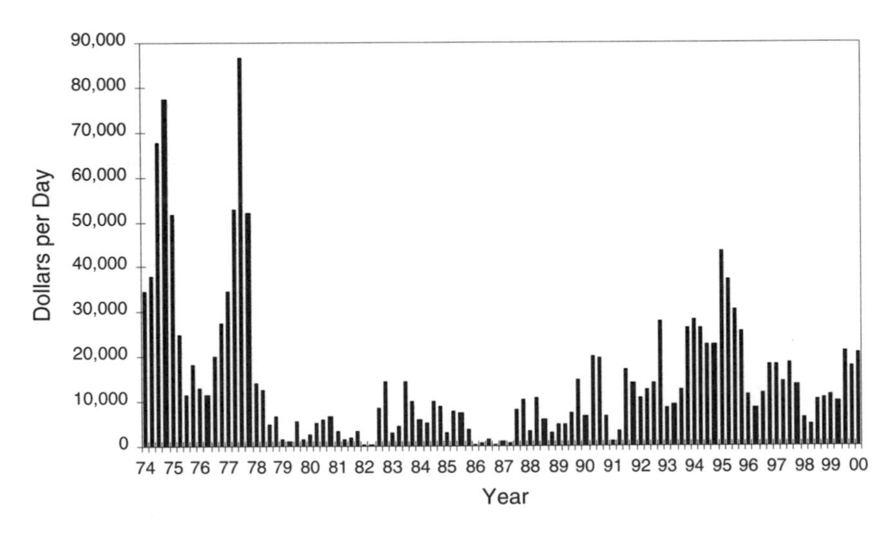

FIGURE G-1 Quarterly average of daily time charter rates for VLCCs operating from the Arabian Gulf to Rotterdam.

CHARACTER OF TANKER SUPPLY

The freight rate volatility in tanker markets is illustrated by the characteristic shape of the supply curve, sometimes referred to as a "hockey stick" (Stopford, 1990). Figure G-2 shows a conceptual representation tracing the tanker capacity that owners are willing to supply at a given freight rate (Hettena and Ruchlin, 1969). Capacity utilization can be expressed as a percentage (as in Figure G-2) or given in cargo tonne-miles or in DWT requirements.

The supply curve is constructed by aggregating the capacity of individual tanker units and arranging them, in ascending order, according to their marginal cost. The marginal cost of a tanker, also called the "lay-up equivalent," is the tanker's operating cost minus the cost of lay-up. For a tanker owner it is the point of economic indifference between keeping the tanker in operation and laying it up. In the short run, the actual revenue falls well below the operating cost because the tanker owner usually resists a costly and disruptive lay-up in the expectation, or hope, that the market will soon recover.

It becomes apparent that the supply curve consists of a number of segments characterized by different elasticities that can be simplified as follows. If demand intersects supply at a point in the neighborhood of *P0,* elasticity is very high: a 5 percent increase in freight rate results in a 25 percent increase in the tonnage supplied. If demand is at *P1,* elasticity is close to unity: a 5 percent increase in freight rate results in a 5 percent increase in the tonnage supplied. If demand increases to *P2,* supply becomes inelastic: a 5 percent increase in freight rates

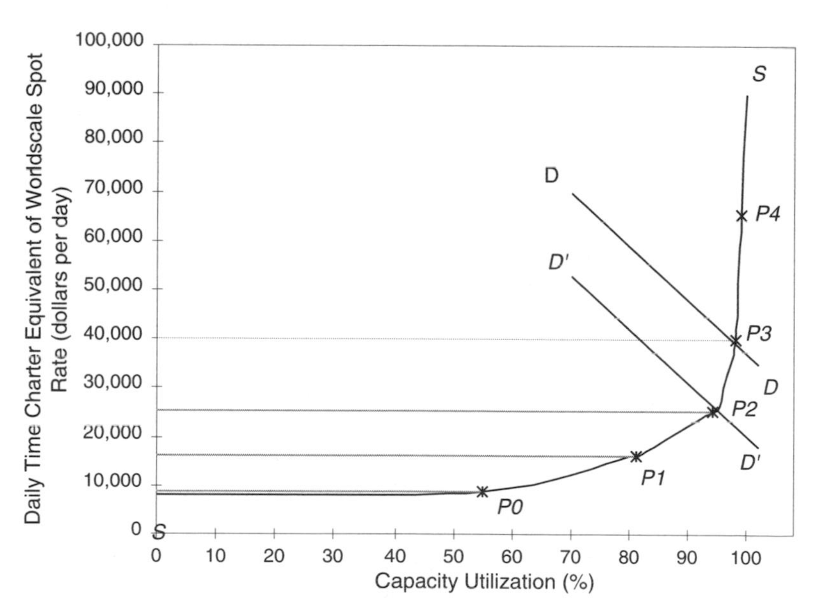

FIGURE G-2 Conceptual VLCC supply function.

results in a 1 percent increase in tonnage supplied. At *P3*, supply becomes highly inelastic: an 87 percent increase in freight rate results in only a 2 percent increase in the tonnage supplied. At *P4* and higher, elasticity is close to zero; no more tankers can be pressed into service; charterers try to outbid each other, and the market turns into an auction. Freight rates can then reach very high levels.

When demand is at *P0*, freight rates are very low and the quantity of tanker tonnage supplied approaches 60 percent of total capacity. At this freight level, only the most efficient tankers operate and they do so at reduced speed, receiving a return at or below operating costs. Under these circumstances, combined carriers operate in the dry bulk trade.

When demand is at *P1,* freight rates are somewhat higher. Many tankers come out of lay-up, and a few combined carriers move to the oil trade. Capacity utilization rises to about 80 percent. Tankers continue to operate at slow speed, and although most tankers cover operating costs, none can cover capital costs.

When demand is at *P2,* freight rates are substantially higher. Almost all tankers come out of lay-up, and more combined carriers move to the oil trade. Most tankers operate at full speed, waiting time is cut down to a minimum, and capacity utilization rises to 95 percent. Most older tankers cover operating and capital costs; newer tankers cover only part of their capital costs.

When demand is at *P3* and higher, freight rates rise to a still higher level. All combined carriers are now in the oil trade, and no further increase in the quantity of tonnage supplied is possible. All tankers operate at full speed, and owners are induced to defer dry-dockings and to reduce downtime to a minimum. The shortage causes many tankers to be utilized inefficiently. Capacity utilization is close to 100 percent, all tankers cover their full costs, and many realize a profit. If this condition continues for an extended period, charterers become concerned about tanker shortages and high freight rates. They seek to enter into long-term contracts and to commit for new tonnage. The tanker fleet can be expected to undergo expansion two or three years hence.

EFFECT OF SEASONALITY OF TANKER DEMAND ON FREIGHT RATES

In contrast to the short-term stability of supply, demand is subject to significant shifts under the impetus of seasonal fluctuations in the oil trade. Seasonal variation in tanker demand has a powerful impact on freight rates. In 1988 and 1989, seasonality accounted for as much as a 10 percent variation in tanker demand, causing VLCC weekly time charter equivalents to vary from about $7,000 per day during the low quarter to about $27,000 per day during the high quarters, as shown in Figure G-3 (Stopford, 1990).

The committee reviewed the degree of seasonal variation in the pattern of oceanborne oil shipments and freight rates in the major oil trades (Gassman, 1996). Seasonal variation indices were computed to identify the periodicity of

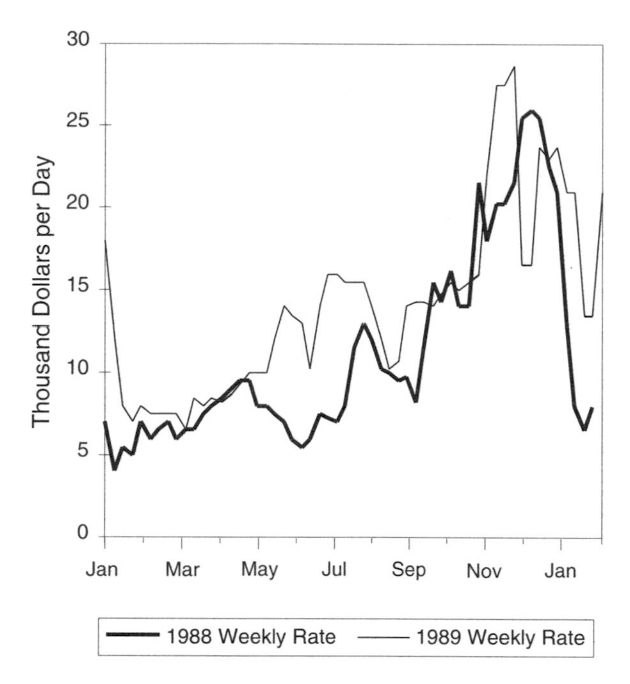

FIGURE G-3 VLCC time charter equivalent rates for 1988 and 1989. Source: Stopford, 1990.

freight rates in different tanker trades. The results of this review are provided in Appendix H.

Aside from seasonality, small shifts in short-term tanker demand are fairly common. As tanker markets approach full capacity utilization, the demand curve DD (Figure G-2) intersects the supply curve at its inelastic segment; hence, small changes in demand have a greatly magnified impact on freight rates. This leads to a condition in which the freight market becomes more and more unstable as demand approaches the limit of capacity. The instability can be precipitated or significantly compounded by seasonal shifts in tanker demand, which have a particularly strong effect on freight rates in periods of tight markets. The market is at its most stable when capacity utilization is low and freight rates are depressed.

REFERENCES

Gassman, W. 1996. Seasonality Trends. Report prepared for the Committee on Oil Pollution Act of 1990 (Section 4115) Implementation Review. Cambridge, Mass.: Massachusetts Institute of Technology, Department of Ocean Engineering.

Hettena, R., and H.S. Ruchlin. 1969. The U.S. tanker industry: A structural and behavioral analysis. Journal of Industrial Economics 18(3):188–204.

Stopford, M. 1990. The supply, demand, and freight rates in the bulk shipping market. Presented at Shipping '90 Conference, Stamford, Connecticut, March 19.

APPENDIX H

Seasonal Variations in Tanker Demand and Freight Rates[1]

The committee has examined and identified the degree of seasonal variation in the pattern of oceanborne oil shipments and freight rates in the major tanker trades.[2]

A monthly seasonal variation index for crude oil exports from the Arabian Gulf (past Hormuz) was computed.[3] The average for the year was defined as 100. A significant degree of seasonal variation is evident from the data presented in Table H-1.

The seasonal index of oceanborne crude oil movements from the Arabian Gulf for the 1990 to 1995 period shows a seasonal high of 102.38 in December and a low of 97.05 in April—that is, a spread of 5.3 percent. The average index for the first five-month period from April through August is 98.21 and for December through March, 101.27. An average seasonal variation of 5 percent in oil shipments is not at all surprising and does not suggest a great seasonal impact on tanker demand. However, if tanker supply is in the inelastic range, the impact of seasonality on freight rates is considerable.

The seasonal variation indices for freight rates of very large crude carriers (VLCCs) trading from the Arabian Gulf to Rotterdam between 1970 and 1995 indicate a pronounced seasonal variation of considerable amplitude (Table H-2). These results support the notion expressed earlier that a small (5 percent) change in tanker shipments can result in a large (47 percent) change in freight rates, consistent with the supply inelasticities discussed in Appendix G.

---

[1]Computations of seasonal indices were performed at the Massachusetts Institute of Technology, Department of Ocean Engineering, Ocean Systems Management, by William Gassman.

[2]Data on crude oil exports were provided by PIRA, April 16, 1996.

[3]Data on freight rates were compiled by Maritime Overseas Corporation based on source material from *Fearnley's Oil and Tanker Markets,* Quarter 4, 1995, and earlier issues.

TABLE H-1 Monthly Indices of Seasonal Variations in Crude Oil Exports, 1990–1995

Month	Oceanborne Crude Oil Exports from Arabian Gulf (past Hormuz)	Crude Oil Exports from Arabian Gulf to Red Sea or Mediterranean	Total Crude Oil Exports from Arabian Gulf
Jan.	101.90	117.10	102.10
Feb.	101.20	138.53	102.20
Mar.	100.90	92.53	101.10
Apr.	97.05	94.22	97.72
May	97.87	92.83	98.66
Jun.	99.16	91.60	99.15
Jul.	99.20	92.42	100.40
Aug.	98.79	91.09	97.63
Sept.	101.40	90.83	99.92
Oct.	100.80	93.86	99.64
Nov.	100.40	102.78	99.66
Dec.	102.38	102.20	101.80
Average	100.00	100.00	100.00

In the case of Suezmax tankers (Table H-3), the same seasonal pattern is present, although somewhat moderated in that the maximum spread of indices is lower. Similar trends are observed for Aframax tankers (Table H-4). For both Suezmax and Aframax tankers, data for the period since 1990 are insufficient to permit conclusions to be drawn.

Data for product tankers trading from the Caribbean to the east coast of the United States are given in Table H-5. Despite the fact that such tankers are smaller and operate in totally different trades from the VLCCs, Suezmax tankers, and Aframax tankers discussed earlier, the impact of the seasonal pattern on freight rates is very strong.

TABLE H-2 Quarterly Indices of Seasonal Variations in Freight Rates of VLCCs Trading from Rotterdam, 1970–1995

Period	Number of Years	Quarter				Average for Year	Maximum Spread between High and Low Indices
		I	II	III	IV		
1970–1995	26	82.99	85.28	116.06	115.67	100.00	33.07
1970–1989	20	72.05	82.64	121.19	124.12	100.00	52.07
1990–1995	6	90.81	72.67	119.67	116.85	100.00	47.00

TABLE H-3 Quarterly Indices of Seasonal Variations in Freight Rates of Suezmax Tankers Trading from the Arabian Gulf to Rotterdam, 1976–1995

Period	Number of Years	Quarter I	Quarter II	Quarter III	Quarter IV	Average for Year	Maximum Spread between High and Low Indices
1976–1994	19	92.54	96.07	103.81	107.58	100.00	15.04
1976–1989	14	77.02	95.70	109.35	117.93	100.00	40.91
1992–1994	3	103.26	91.71	95.11	109.92	100.00	18.21

TABLE H-4 Quarterly Indices of Seasonal Variations in Freight Rates of Aframax Tankers Trading from North Africa to Rotterdam, 1976–1995

Period	Number of Years	Quarter I	Quarter II	Quarter III	Quarter IV	Average for Year	Maximum Spread between High and Low Indices
1976–1995	20	95.63	94.86	94.93	114.59	100.00	19.73
1976–1989	14	85.98	91.05	98.18	124.79	100.00	38.81
1992–1995	4	98.46	99.23	94.24	108.07	100.00	13.83

TABLE H-5 Quarterly Indices of Seasonal Variations in Freight Rates of Product Tankers Trading from the Caribbean to the U.S. East Coast, 1976–1995

Period	Number of Years	Quarter I	Quarter II	Quarter III	Quarter IV	Average for Year	Maximum Spread between High and Low Indices
1976–1995	20	113.27	84.72	86.60	115.60	100.00	30.88
1976–1989	14	108.08	84.84	90.37	116.71	100.00	31.87
1992–1995	4	118.18	95.13	87.08	99.62	100.00	31.10

APPENDIX I

Letter to the Committee from R.W. Porter

LONDON TANKER BROKERS' PANEL LIMITED

Directors:
E. F. Shawyer (Chairman)
M. G. Johnson
A. G. Burgess
R. W. Park
C. Waaler
I. M. Shaw
R. W. Porter (Managing)

Prince Rupert House
64 Queen Street
LONDON EC4R 1AD

Telephone: 0171-248 4747
Telex: 885118 G
Fax: 0171-489 0536

Our Ref: RWP/EAL
Your Ref: MB-96-528

25th April 1996

Mr. D. Wolcott
Chair, Committee on OPA 90 (Sn 4115)
Marine Board
National Research Council
2101 Constitution Avenue
Washington D.C. 20418

Dear Mr. Wolcott

We thank you for your letter to Mr. Eric Shawyer, Chairman of the London Tanker Brokers' Panel Limited of 5th April 1996 regarding evaluating various aspects of the international tanker market. Your letter has been distributed to all Members of the LTBP and discussed at length at a recent meeting of the LTBP. The Panel Members have asked me to reply on their behalf.

The research required to attempt to evaluate any variations in the world tanker market for the specific points you require would be extremely extensive in nature and unlikely to provide meaningful results as there is no obvious differentiation on the basis of hull type, age etc, in market negotiations with charterers taking decisions on the acceptability of individual vessels on a wide variety of bases backed up by varying degrees of physical inspection and documentary investigation. Anecdotal evidence would suggest that in an otherwise free choice, a charter would choose to utilise the youngest, most technically sound vessel available whilst paying the same "market" freight rate applicable to all the vessels available.

A recent regional variation of note illustrates this point in that whilst Korean authorities have imposed a 20 year age limit on vessels entering their water there is no definite evidence that this has affected the "market freight rates applicable to such voyages.

In summary it is true that there are suggestions that hull type, age etc, influences the freight rates paid in particular markets such variations, if they exist, are not apparent in the available market reports. It is not clear whether the significant resources which would be required to be allocated to this sort of investigation would provide any useful or meaningful results given the above reservations.

On a more general note it has always been the LTBP's policy not to express an opinion on tanker market matters but to provide specific responses, based on the available market information, to a specific query from a client and Members would be reluctant to alter this situation. It may well be the case that the individual companies which provide expertise to the Panel may not share this reticence.

We regret that we are unable to give definitive answers to your queries but hope that the above information is of some assistance to your committee.

Please do not hesitate to ask if you feel there is something the Panel could assist with.

Yours sincerely
LONDON TANKER BROKERS' PANEL LIMITED

R.W. Porter

R.W. Porter
Managing Director

APPENDIX

J

Methodology for Determining Ship Replacement Costs

This appendix describes the methodology used in Chapters 4 and 5 in calculating whether a tanker owner will choose to extend the life of an existing single-hull tanker or replace it with a new double-hull tanker. The methodology is adapted from ICF Kaiser (1995). Data used in the committee's analysis are provided in the tables.

The analysis simplifies the decision process by assuming that the only decision to be made is whether or not to go through the fifth special survey[1,2] and operate for another five years, after which the tanker will be scrapped and replaced. The annualized cost of continuing to operate the single-hull vessel, including the special survey costs, is compared with the annualized cost of purchasing and operating a new double-hull vessel in place of the single-hull vessel.

The least expensive alternative for a pre-MARPOL (the International Convention for the Prevention of Pollution from Ships, adopted in 1973 and amended in 1978) tanker when operating after age 25 is hydrostatically balanced loading (HBL), although such a tanker cannot trade to the United States unless it uses the deepwater port or the lightering areas. No investment is required, but cargo capacity is reduced between 3 and 8 percent because of the HBL requirement. To operate the tanker another five years, the annual costs involved are:

$$EC/year = OC_{hbl} + \{CAP_{hbl} - [SCRAP/(1 + i)^5] + SS\} \times \{i/[1 - (1 + i)^{-5}]\}$$

---

[1]Special surveys are required every five years by classification societies and will apply whether the shipowner is operating under the International Maritime Organization Regulation 13G or the Oil Pollution Act of 1990 (P.L. 101-380).

[2]For the Jones Act vessel analysis in Chapter 5, the methodology has been modified to include decision points at both the fifth and the sixth special surveys.

EC/year is the annualized extension cost of passing the special survey and operating another 5 years. OC_{hbl} is the operating cost of the tankers with reduced capacity due to HBL, *SS* is the special survey cost. CAP_{hbl} is the capital cost associated with HBL, assumed to be zero. *i* is the annual discount rate assumed to be 10 percent. $[1 - (1 + i)^{-5}]$ is a factor that spreads out capital costs into their annualized equivalents, using the discount rate. *SCRAP* is the value of the tanker when scrapped. Each year the vessel is kept in operation delays realization of the scrap value by one year, so the scrap value is divided by a discount factor raised to a power equal to the number of years of life remaining for the tanker, assumed to be 5.

The costs of operating the tanker must be compared to the costs of buying a new tanker. This comparison must include a factor reflecting the reduced carrying capacity of the existing tanker with HBL. The annual costs of a new double-hull tanker are:

$$NC/year = \{OC_{new} + (CAP_{new}) \times [1 - (1 + i)^{-25}]\} \times (1 - CAPREDUC)$$

NC/year is the annualized cost of buying and operating a new double-hull vessel, as adjusted for changes in the carrying capacity of the existing vessel. OC_{new} is the annual operating cost for the new tanker. *i* is the annual discount rate, assumed to be 10 percent. $[1 - (1 + i)^{-25}]$ is a factor that spreads out capital costs into their annualized equivalents over 25 years (the assumed economic life in years for a new double-hull tanker) using a discount rate *i*. CAP_{new} is the capital cost of the new tanker. *CAPREDUC* is the reduction in carrying capacity in an existing vessel as a result of HBL (i.e., reducing the costs of the new vessel to compare it with the decreased carrying capacity of the existing tanker).

The most difficult factor to estimate is generally the special survey cost SS. For an excellently maintained vessel the minimum cost will be $435,000 for dry-docking and survey (with no extra costs for repairs). Typically, one would expect the cost to be several million dollars more if repairs such as replacement of steel and the opportunity cost of an extended stay in the repair yard are included. The equations above can be evaluated using estimated data. If the special survey cost is not known, the equations can be solved for it as shown below.

$$\begin{aligned} SS = & \left(OC_{new} + (CAP_{new}) \times \{i/[1 - (1 + i)^{-25})]\}\right) \\ & \times \left((1 - CAPREDUC)/\{i/[1 - (1 + i)^{-5}]\}\right) \\ & - OC_{hbl}/\{i/[1 - (1 + i)^{-5}]\} \\ & - [CAP_{hbl} - SCRAP/(1 + i)^{5}] \end{aligned}$$

This value can be expressed as the break-even special survey cost. If the actual

TABLE J-1 Data for Calculating the Cost of Tankers in International Trade ($ million)

Vessel Size (DWT)	Existing Vessel Parameters					Double-Hull Parameters		
	OC_{hbl}	CAP_{hbl}	*SCRAP*	*CAPREDUC (%)*	*i (%)*	OC_{new}	CAP_{new}	*i (%)*
40,000	3.80	0	1.60	0, 5, 8	10.0	3.20	33.50	10.0
60,000	4.10	0	1.80	0, 5, 8	10.0	3.40	37.00	10.0
140,000	5.10	0	3.20	0, 5, 8	10.0	4.20	54.00	10.0
280,000	7.00	0	5.20	0, 5, 8	10.0	5.80	85.00	10.0

special survey cost is higher than this, buying a new double-hull tanker would cost less than trying to get another five years out of the old single-hull tanker. Alternatively, if the actual special survey cost is lower than the break-even special survey cost, the owner would opt to continue operating the single-hull tanker for another five years.

Data used in cost calculations for the international trade in Chapter 4 and the Jones Act trade in Chapter 5 are provided below in Tables J-1 and J-2, respectively. The cost data were taken from the ICF Kaiser study (1995) and other sources used in the report. Although vessels are presented as being particular sizes, each ship actually represents a size range.

TABLE J-2 Data for Calculating Costs for Jones Act Tankers ($ million)

Vessel Size (DWT)	Existing Vessel Parameters					Double-Hull Parameters		
	OC_{hbl}	CAP_{hbl}	SCRAP	CAPREDUC (%)	i (%)	OC_{new}	CAP_{new}	i (%)
40,000 (tanker)	7.60	0	1.60	0, 10, 24	10.0	5.60	41.90	10.0
120,000 (tanker)	11.20	0	2.80	0, 10, 24	10.0	8.10	67.00	10.0
13,000 (barge)	3.20	0	1.10	0, 10, 24	10.0	3.20	9.60	10.0

REFERENCE

ICF Kaiser. 1995. Regulatory Assessment of Supplemental Notice of Proposed Rulemaking on Structural Measures of Existing Single-Hull Tankers. Prepared for U.S. Department of Transportation. Cambridge, Mass.: Volpe National Transportation System Center.

APPENDIX K

Comparative Study of Double-Hull and Single-Hull Tankers[1]

INTRODUCTION

Prior to 1990, most crude oil carriers were built with single hulls. Design, construction, and operational experience of double-hull tankers was limited primarily to product and parcel tankers under 40,000 tons deadweight. The stability and strength characteristics of double-hull crude oil carriers are quite different from single-hull tankers and product carriers, and designers and operators of double-hull tankers found themselves confronted with a new set of issues to consider.

This appendix examines the design characteristics of double-hull tankers built since 1990. Four of the areas in which double-hull tankers perform differently as compared to single-hull tankers have been identified and investigated. These are:

- environmental performance with regard to oil outflow from collisions and grounding
- survivability characteristics after experiencing a collision or grounding
- intact stability during load and discharge operations
- hull girder strength and draft considerations for the ballast condition

For comparative purposes, both single-hull and double-hull configurations have been investigated. Double-hull ships are selected to be representative of the tankage arrangements and proportions typically built since 1990. The size of a tanker has a significant influence on the stability and survivability characteristics of the vessel, and therefore the designs studied are divided into the following five groups:

---

[1]Prepared for the Committee on Oil Pollution Act of 1990 (Section 4115) Implementation Review by Herbert Engineering Corporation, San Francisco, California, April 15, 1996.

- tankers of 35,000 DWT–50,000 DWT
- tankers of 80,000 DWT–100,000 DWT
- tankers of 135,000 DWT–160,000 DWT
- tankers of 265,000 DWT–300,000 DWT
- oceangoing barges

SUBDIVISION NOMENCLATURE

The following terms are used to describe the ship's subdivision:

- *Cargo block.* The cargo block is the portion of the ship extending from the forward boundary of the forward-most cargo tank to the aft boundary of the aft-most cargo tank. OPA '90 as well as the 1992 Amendments to Annex I of MARPOL 73/78 require that all oil tanks within this space be segregated from the side and bottom shell.
- *Cargo tanks.* All tanks arranged for the carriage of cargo oil. Unless noted otherwise, the term "cargo tanks" shall be assumed to include the slop tanks.
- *Slop tanks.* Slop tanks are provided for storage of dirty ballast residue and tank washings from the cargo tanks. Annex I of MARPOL 73/78 requires that tankers be arranged with slop tanks.
- *Cargo tank arrangements.* Figure K-1 shows cross-sections of typical cargo tank arrangements for double-hull tankers. The "STA" or single-tank-across arrangement has a single center cargo tank spanning between wing tanks. This design is frequently arranged with upper hopper tanks in way of the outboard wings, in order to reduce the free surface when the cargo tanks are nearly full. The two-tanks-across arrangement has a centerline bulkhead and port and starboard cargo tanks. Vessels under 160,000 DWT are typically arranged as single tank across, two tanks across, or a combination thereof. Most larger tankers are arranged with three tanks across as required to satisfy the MARPOL requirements for tank size and damage stability.

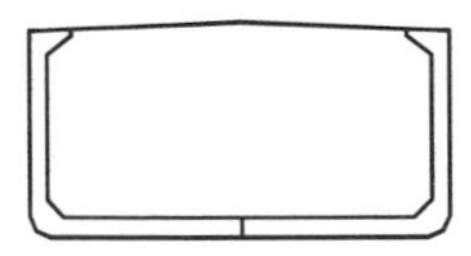
Single Tank Across

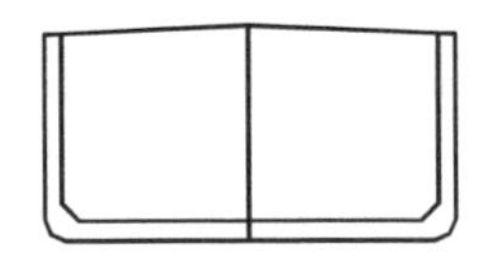
Two Tanks Across

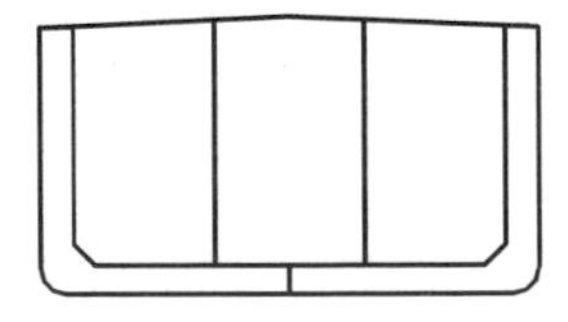
Three Tanks Across

FIGURE K-1 Cargo tank arrangements.

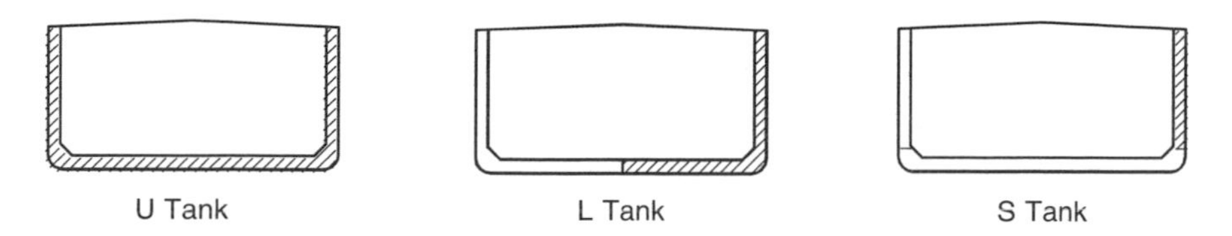

FIGURE K-2 Ballast tank arrangements.

- *Ballast tank arrangements.* Figure K-2 shows typical ballast tank configurations.
  - "L" tanks are the most commonly used configuration. L tanks are usually aligned with the cargo tanks, although they will occasionally extend longitudinally over two cargo tanks.
  - "U" tanks reduce asymmetrical flooding, and are generally used when L tank arrangements fail to meet damage stability requirements. U tanks extend over the full breadth of the ship, and have a significantly higher free surface as compared to a pair of L tanks.
  - "S" or side tanks are located entirely in the wing tanks. S tanks improve the survivability characteristics of a vessel as they normally will not be penetrated when bottom damage is incurred.

METHODOLOGY AND ASSUMPTIONS

Oil outflow, survivability, intact stability, ballast draft, and strength evaluations have been carried out for 27 tankers. These are all vessels that have either been delivered or are currently under contract. Oil outflow and survivability calculations have also been carried out for nine barges. All calculations have been done using HECSALV (Herbert Engineering Corporation, 1996) software. The calculation methodology and assumptions are described below.

Evaluating Oil Outflow

All cargo oil tanks on a double-hull tanker built to OPA 90 requirements are protectively located. Many of the damage cases that would result in oil spillage on single-hull tankers will not penetrate the cargo tanks of double-hull tankers. Double-hull tankers will have fewer accidents involving oil spillage. The mean or expected oil outflow from a casualty will usually be less with a double-hull tanker as compared to a single-hull tanker of the same size.

The arrangements of double-hull tankers vary. The vessel proportions, the wing tank and double bottom dimensions, and the number and location of longitudinal and transverse bulkheads all influence the outflow performance. As a consequence, the likelihood of oil spillage and the mean or expected oil outflow will vary significantly even among double-hull tankers of the same size.

The International Maritime Organization (IMO) guidelines (1995) for evaluating alternatives to double-hull tankers have been applied in this report for assessing oil outflow performance. Although originally intended for evaluating alternatives to the double-hull concept, these guidelines are also well suited for comparing the outflow performance of single-hull and double-hull tankers. The guidelines take a probabilistic approach based on historical statistical data, and provide a methodology for assessing both the likelihood of a spill and the expected outflow. The IMO guidelines account for factors such as varying wing tank widths and double bottom heights, the influence of internal subdivision, the effects of tide, and the influence of dynamic effects on outflow.

Principles of Oil Outflow

The following provides a brief description of the fundamental principles affecting oil outflow. More extensive discussions are contained in *Tanker Spills, Prevention by Design* (NRC, 1991) and the USCG report, *Probabilistic Oil Outflow Analysis of Alternative Tanker Designs* (DOT, 1992).

Hydrostatic Balance. In the event of bottom damage, oil outflow will occur until the internal pressure exerted by the entrapped oil and flooded water within a tank equals the external pressure exerted by the seawater. If the ullage space is underpressurized such that the pressure on the oil surface is less than the atmospheric pressure acting on the seawater, outflow will be reduced. Conversely, higher ullage space pressures as might be introduced by the inert gas system will result in larger outflows. For groundings, the external pressure is reduced as the tide drops, and outflow will occur until equilibrium is once again attained.

For lightly loaded tanks, the initial pressure head from the cargo oil is less than the external seawater pressure. When bottom damage is sustained, seawater enters the bottom of the tank until equilibrium is achieved. Provided the damage does not extend up the side of the tank and currents or vessel motions do not induce mixing of seawater and oil in the vicinity of the damage, no oil will be lost.

Oil Entrapment in Double-Hull Tankers. When a tanker experiences bottom damage through the double bottom tanks and into the cargo tanks, a certain portion of the oil outflow from the cargo tanks will be entrapped by the double bottom tanks. This phenomenon was investigated through model testing at the David Taylor Research Center (DTRC, 1992) and the Tsukuba Institute, Ship & Ocean Foundation (Tsukuba Institute, 1992), and through numerical analysis. These studies indicate that oil entrapment is influenced by many factors, including the size and location of openings, the magnitude of the pressure imbalance, and whether the double bottom tank is flooded with water at the time the oil tank is ruptured. For conditions in which the double bottom initially floods and then the cargo tank is breached, a viscous jet is formed resulting in minimal retention of oil in the outer hull. The Marine Environmental Protection Committee (MEPC) concluded that "if both outer and inner bottoms are breached simultaneously and the extent of rupture at both bottoms is the same, it is probable that the amount of sea water and oil flowing into the double-hull space would be the same." In its regulations, IMO assumes that double bottoms below oil tanks retain a 50:50 ratio of oil to sea water. Where tidal changes introduce a slowly changing pressure differential, higher retention rates can be expected.

Dynamic Oil Losses. Oil losses in excess of those predicted from hydrostatic balance calculations may result due to the initial impact when a vessel runs aground, and subsequently, from the effects of current and ship motions. These losses primarily influence single-hull vessels and alternative designs whose oil tanks contact the outer hull.

Model tests at David Taylor Research Center (1992) and the Tsukuba Institute (1992) were carried out to assess the influence of initial impact and current on oil outflow. Dynamic losses are influenced by the speed of the ship, the extent of damage, the magnitude of the current, and the sea state. Under extreme weather conditions, losses up to 10 percent of the tank volume can be encountered, although dynamic oil losses of 1 percent to 2 percent are more typical. In its regulations, IMO assumes a minimum outflow of 1 percent of the volume for all breached cargo tanks which bound the outer hull.

Side Damage. The location and size of the damage opening influences the amount of expected oil outflow from side collisions. If the lower edge of the damage opening lies above the equilibrium waterline, the oil level in the tank will drop to the height of the opening and the vessel will heel away from the damage.

When the damage extends below the waterline, outflow of oil will occur until hydrostatic balance is achieved. Over time, all oil located below the level of the upper edge of the damage opening will be replaced by the denser seawater. In its regulations, IMO assumes that 100 percent of the oil in breached side tanks is lost.

Methodology for Evaluating Oil Outflow

Each of the designs has been evaluated using the conceptual analysis approach (without consideration of survivability) as defined in the IMO Interim Guidelines for Approval of Alternative Methods of Design and Construction of Oil Tankers under Regulation 13F(5) of Annex I of MARPOL 73/78 (IMO, 1995). An overview of the methodology is described below. Further details on application of these regulations can be found in Michel and Moore (1995).

The IMO guidelines call for the calculation of three parameters: the probability of zero outflow, mean outflow, and extreme outflow. The calculation methodology assumes the vessel experiences a collision or grounding, and that the outer hull is breached. The assumed extent of penetration, and therefore the probability that the inner hull of a double-hull tanker will be pierced, are based on the application of probability density functions as described in the following paragraphs.

The probability of zero outflow is the likelihood that such an encounter will result in no cargo oil spillage into the environment, and is an indicator of a design's tendency towards avoiding oil spills. The mean outflow is the weighted average of the cumulative oil outflow, and represents the expected or average outflow. This mean outflow provides an indication of a design's effectiveness in mitigating the amount of oil loss due to collisions and groundings. The extreme outflow is the weighted average for the most severe damage cases, and provides an indication of a design's effectiveness in reducing the number and size of large spills.

Historical data from collisions and groundings of tankers were collected by a number of classification societies under the direction of IMO (Lloyds Register of Shipping, 1991), and reduced into probability density distribution functions. The area under the probability density curve between two points on the horizontal axis is the probability that the quantity will fall within that range. The density distribution scales are normalized by ship length for location and longitudinal extent, by ship breadth for transverse location and transverse extent, and by ship depth for vertical location and vertical extent. Statistics for location, extent, and penetration are developed separately for side and bottom damage cases.

Figure K-3 shows the probability density distribution for the longitudinal extent of grounding damage. The histogram bars represent the data collected by the classification societies, and the linear plot represents IMO's piece-wise linear fit of the data. The area under the curve up to a damage length/ship length of 0.3 equals 0.75. Based on these statistics, there is a 75 percent likelihood that the longitudinal extent of damage for a ship involved in a grounding incident will not exceed 30 percent of the ship's length.

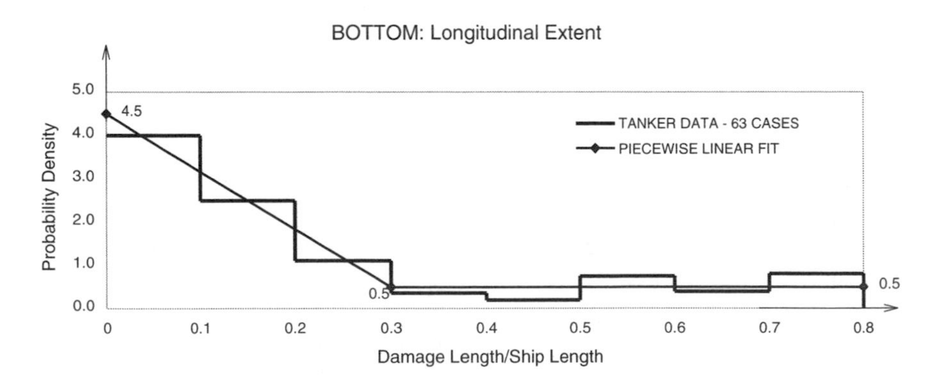

FIGURE K-3 Longitudinal extent of grounding damage.

Through application of these functions to the hull and compartmentation of a particular vessel, all possible combinations of damaged compartments are determined, together with their associated probabilities of occurrence. Calculations are then performed to determine the oil outflow associated with each of these incidents. For the vessels analyzed in this study, the number of unique damage cases ranged between 100 and 350 for side damage, and between 300 and 700 for bottom damage.

For side damage incidents, 100 percent oil loss is assumed for each breached cargo tank. Therefore, if a given damage incident damages only a ballast wing tank, zero outflow occurs. If a damage incident involves breaching of the ballast wing tank and the adjacent cargo oil tank, the full contents of the cargo oil tank are assumed to be lost.

For bottom damage, outflow is determined by performing hydrostatic pressure balance calculations. A reduction in tide after the incident of 0.0 meters, 2.0 meters, and 6.0 meters (or one-half the draft, whichever is less) is assumed. Other assumptions applicable to bottom damage calculations are:

- An inert gas pressure of 0.05 bar is applied to all cargo oil tanks. This is a positive pressure and augments the oil outflow.
- If a double bottom ballast tank or void space is located immediately below a breached cargo tank, the flooded volume of the double bottom tank is assumed to be a 50:50 mixture of oil and seawater. The oil entrapped in the double bottom is not included in the assumed spill volume.
- For breached cargo tanks bounding the bottom shell, oil outflow equal to 1 percent of the tank volume is assumed as the minimum outflow. For tanks which are hydrostatically balanced in the intact condition, outflow analysis based on hydrostatic-balance principles will indicate zero outflow for grounding cases not subject to tidal change. In these circumstances, the minimum outflow value accounts for oil loss due to initial impact and the effects of current and waves.

Independent calculations are carried out for side and bottom damage, and the three outflow parameters computed. For the grounding evaluation, the 0.0 meter, 2.0 meter, and 6.0 meter tidal change results are combined in a 40 percent:50 percent:10 percent ratio. The side and bottom damage results are then combined in a 40 percent:60 percent ratio. A pollution prevention index is developed by substituting the outflow parameters for the actual design and the IMO reference double-hull design into the following formula provided in the IMO Guidelines. If the Index E is greater than or equal to 1.0, the alternative design is considered at least equivalent to the IMO reference design.

$$E = \frac{(0.5)(P_0)}{P_{OR}} + \frac{(0.4)(0.01 + O_{MR})}{0.01 + O_M} + \frac{(0.1)(0.025 + O_{ER})}{0.025 + O_E}$$

P_0 = probability of zero outflow for the alternative design. O_M = mean oil outflow parameter for the alternative design = (mean outflow)/C. O_E = extreme oil outflow parameter for the alternative design = (extreme outflow)/C. C = total cargo oil onboard. P_{OR}, O_{MR}, and O_{ER} are the corresponding parameters for the reference double-hull design of the same cargo oil capacity.

The IMO reference double hulls are shown in Figure K-4. These reference designs do not represent the minimum subdivision acceptable under current MARPOL regulations. Rather, it was IMO's intent to select designs which "exhibit a favorable oil outflow performance." For instance, the 150,000 DWT reference ship has a 6 × 2 cargo tank arrangement, whereas a 5 × 2 arrangement is permissible under current rules. Similarly, the assumed double bottom depth on the VLCC is in excess of the rule requirements.

The IMO Guidelines specify that C, the cargo oil onboard, be taken at 98 percent of the total cargo tank volume, and that the density of the cargo oil be as required to bring the vessel to its subdivision draft. For this analysis, it is assumed that each vessel is loaded to its summer load line with crude oil at a density of 0.90 metric tons/m^3. This typically means that one tank or pair of tanks is partially full. The partially loaded tank or tanks were selected in order to maintain a trim in the intact condition between zero and 0.5 meters by the stern. In all other respects, the analysis has been carried out in strict conformance with the IMO guidelines.

Survivability Evaluation

Most single-hull tankers have excellent damage stability characteristics. When cargo oil tanks are breached, the oil is displaced by seawater of comparable or slightly higher density, resulting in relatively small heeling moments. For MARPOL 78 tankers, the side ballast tanks will introduce an asymmetric heeling moment. However, these tanks are arranged adjacent to cargo tanks. MARPOL 78 tankers are designed to withstand damage to a ballast tank, or to the ballast tank and an adjacent cargo tank. Breaching two ballast tanks would require damage extents longer than the length of a cargo tank, and the probability of such extents is extremely small.

Double-hull tankers are arranged with wing ballast tanks along the length of the cargo block. When breached, these tanks introduce asymmetric loading which will tend to heel the vessel in the direction of the damage. In addition, the double bottom raises the height of the cargo oil, which translates into a higher center of gravity for the intact condition as compared to a single-hull tanker. Free surface effects may also be higher, as single-tank-across arrangements of cargo tanks are not uncommon in double-hull tankers. These effects all tend to increase the heeling moment. Excessive asymmetrical flooding will lead to immersion of down flooding points, and eventually the vessel will sink or capsize.

IMO recognized the potential survivability problems with double-hull tankers.

In Regulation 13F of the 1992 Amendments to Annex I of MARPOL 73/78 (IMO, 1992), the two compartment damage stability criterion contained in Regulation 25 of Annex I of MARPOL 73/78 was supplemented with raking bottom damage requirements.

Regulation 13G and Regulation 25 both use a deterministic analysis approach in which fixed damage extents are assumed. Such calculations do not provide a clear picture of the survivability characteristics of a vessel. In this report, survivability is evaluated by applying the probabilistic density distribution functions for side damage as contained in IMO guidelines (IMO, 1995) for evaluating alternative tanker designs together with the damage survival requirements defined in Regulation 25.

Methodology for Evaluating Survivability

The principles affecting damage stability and survivability calculations are well documented in the literature (SNAME, 1988; IMO, 1993). The vessel is assumed to sustain damage which breaches the outer hull. Damaged compartments are assumed to be in free communication with the sea. The vessel sinks lower, trims, and heels until equilibrium is reached.

A reiterative calculation approach is applied to determine the equilibrium draft and trim conditions over a range of heel angles. The computed heeling moment at each angle is then divided by the original intact displacement of the vessel less any fluid outflow, in order to develop the righting arm or "GZ" curve. From the GZ curve, the equilibrium heel angle can be determined. Properties of the GZ curve, such as its maximum value, positive range, and the area under the curve provide an indication of the reserve stability of the damaged vessel.

Current analytical techniques do not provide a means for accurately determining the probability that a damaged ship will not capsize or sink. The assessment of survival or non-survival for a given damage case is therefore done on a deterministic basis. For instance, the IMO damage stability criteria for passenger ships, dry cargo ships, and tankers all contain minimum requirements regarding immersion of down flooding points, maximum heel angles, and residual stability. When these values are attained, survival is assumed. It is generally recognized that the IMO criteria reflect survival rates in a relatively moderate sea state, perhaps Beaufort force 3 or 4.

For this study, the probability of flooding each combination of compartments has been determined from the probability density functions defined in the IMO guidelines. Only side damage from collisions has been considered when evaluating survivability.

The vessel is assumed to be fully loaded to the summer load line draft. Consumables are assumed to be 50 percent full, and all cargo tanks 98 percent full. Where breached tanks are filled or partially filled, it is assumed that 100 percent of the fluid in the tank is displaced by seawater.

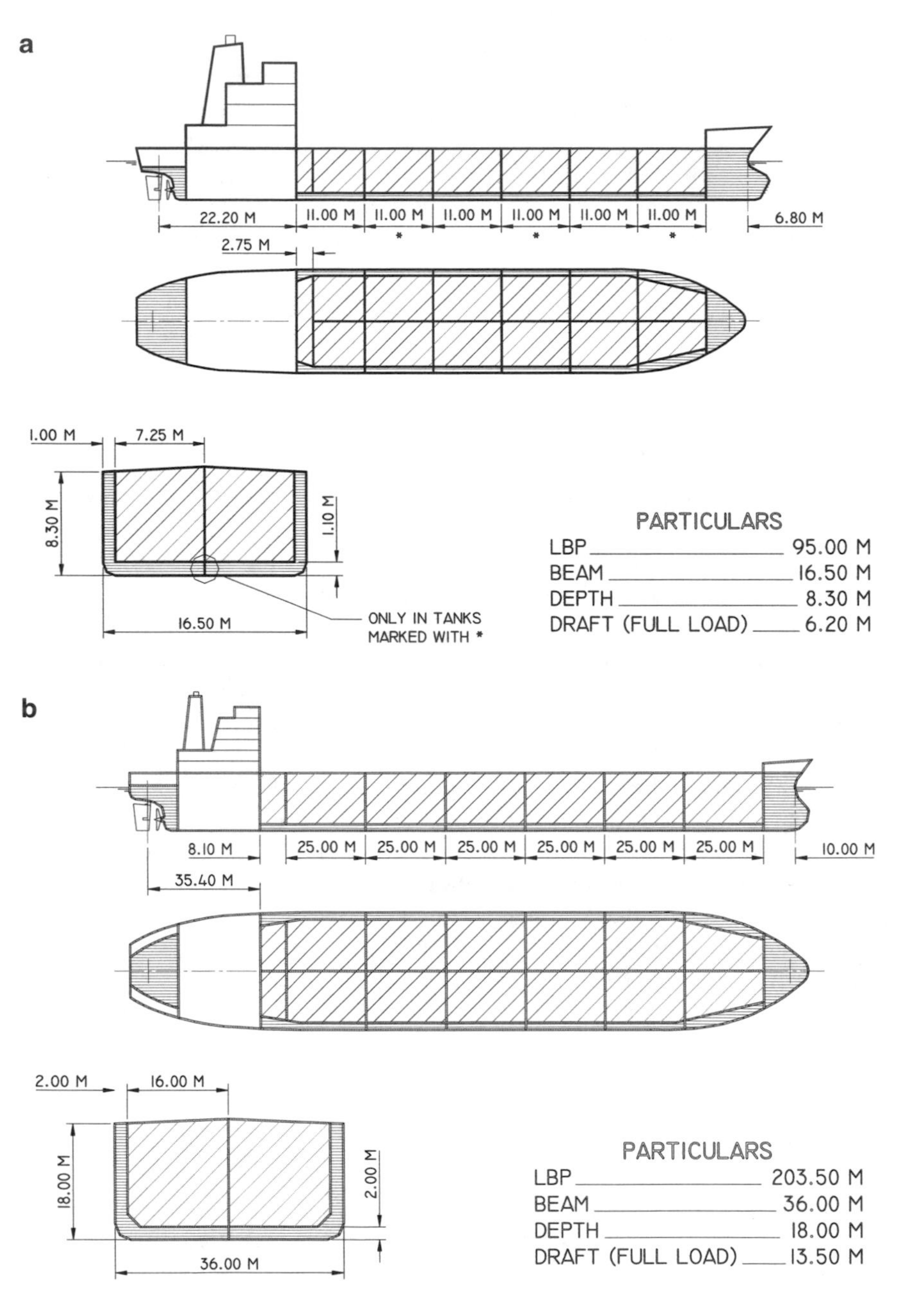

FIGURE K-4 IMO reference double hulls. (a) IMO Double-Hull Reference Design No.1, 5,000 DWT. (b) IMO Double-Hull Reference Design No.2, 60,000 DWT. (c) IMO Double-Hull Reference Design No.3, 150,000 DWT. (d) IMO Double-Hull Reference Design No.4, 283,000 DWT.

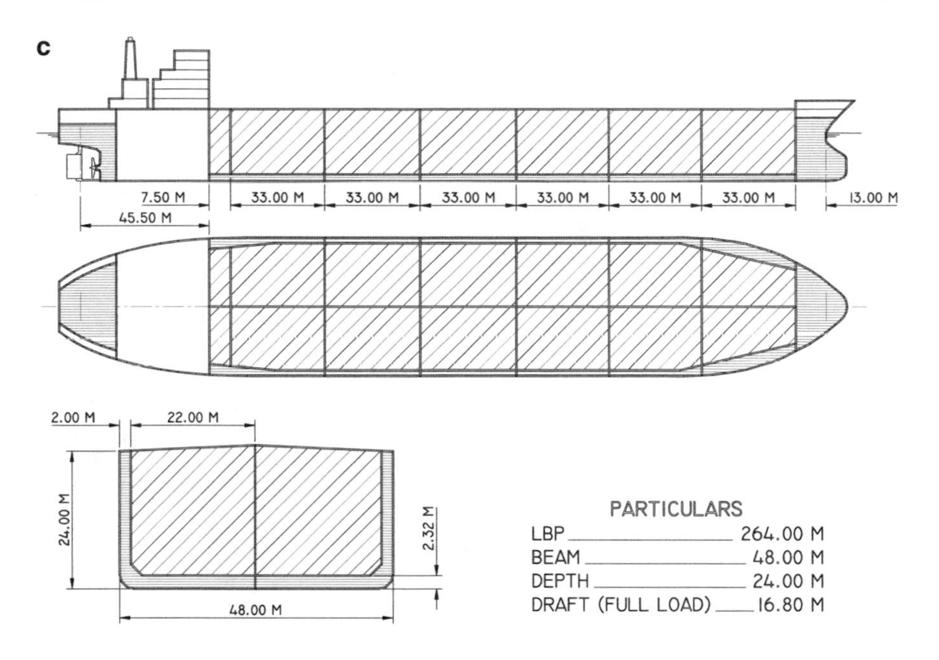

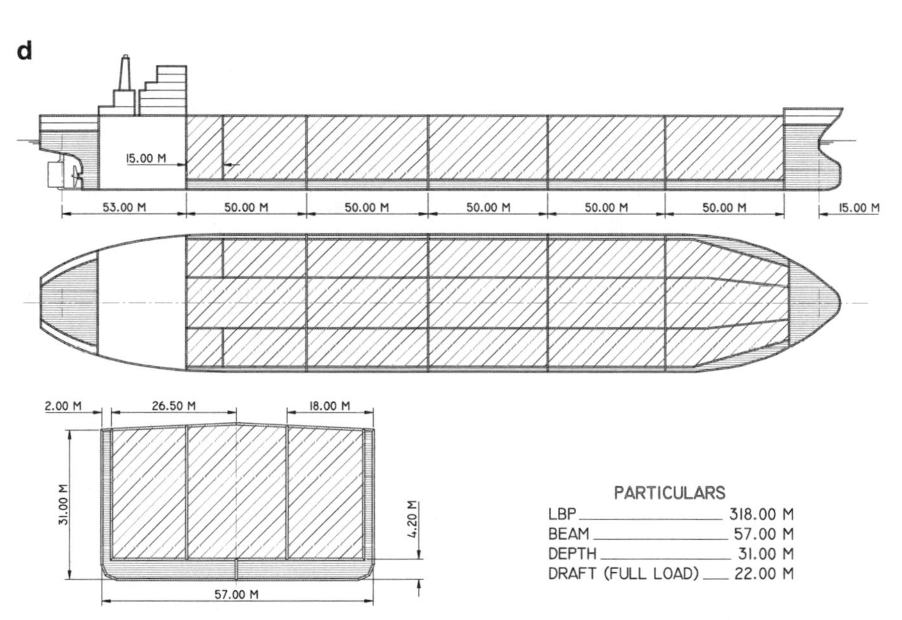

FIGURE K-4 *Continued*

The assessment of survivability is based on a comparison with the IMO regulation 25 (3) of Annex I of MARPOL 73/78. These limits are as follows:

- *Equilibrium heel angle.* Maximum 30 degrees, or 25 degrees if the deck edge is immersed.
- *Righting arm.* Maximum residual righting lever of at least 0.1 meters.
- *Range of positive stability.* Range of positive stability beyond the equilibrium heel angle of at least 20 degrees.
- *Progressive flooding.* Downflooding points such as overflows and air pipes for all nonbreached compartments shall not be immersed at the equilibrium waterline.

An index of survivability has been determined by summing the probabilities for each damage case which satisfies these survival criteria. Typically, index values fall between 97 percent and 100 percent.

Intact Stability Evaluation

Single-hull tankers are inherently stable. The MARPOL regulations for hypothetical outflow, tank length, and damage stability dictate the tank size, and tend to encourage an arrangement of the longitudinal bulkheads such that wing tanks and center tanks have comparable widths. Furthermore, single-hull tankers built to MARPOL 73 and MARPOL 78 requirements typically have only two and four ballast tanks, respectively, within the cargo block. These ships have relatively small free surface effects, even when all cargo and ballast tanks are slack simultaneously. Since it is not possible to create an unstable situation for most single-hull tankers, IMO did not institute intact stability requirements for tankers.

In contrast, double-hull tankers have ballast tanks covering the entire cargo block. Structural and cost optimization under current MARPOL regulations tend to encourage larger tanks and a minimization of longitudinal bulkheads. For tankers under 120,000 tons deadweight, the low-cost solution is to have minimum wing tank widths (1 to 2 meters), with single cargo tanks spanning between wing tank bulkheads. The increase in the number of ballast tanks and the tendency towards wider ballast and cargo oil tanks means increased free surface effects, and a reduction in stability. This reduction in stability is exacerbated by the rise in the center of gravity of the cargo oil due to the double bottoms. As a result, some double-hull designs are unstable for certain combinations of ballast and cargo loading. There have been a number of incidents in the last few years in which tankers have become unstable during cargo operations. Although no tankers have capsized at the pier, angles of loll up to 15 degrees have been reported.

Principles of Intact Stability

The stability of a ship is influenced by a number of factors: the vertical center of gravity of the ship, the free surface of liquids within tanks, and the righting moment developed as the vessel heels.

The vessel shown in Figure K-5 (a) exhibits positive transverse stability. As the vessel heels, the center of buoyancy shifts from B to B1. The buoyancy force acts upward through the center of buoyancy B1, and the weight of the vessel acts downward through the vertical center of gravity G. The distance GZ is the righting arm. As the buoyancy force is tending to right the vessel, the ship is stable, and the righting arm GZ is positive.

The vessel in Figure K-5 (b) illustrates the impact of the rise in the center of gravity on stability. The heeling moment has increased to where it now exceeds the buoyancy moment, and the vessel has negative stability. The weight force is now acting outboard of the buoyancy force, and the righting arm GZ is negative.

As the vessel heels, liquids in partially full tanks shift towards the low side. This moves G in the direction of heel, reducing the righting arm GZ. This phenomenon is called the free surface effect. For a rectangular tank, the free surface varies as the cube of the width of the tank.

Figure K-6 shows typical tanker designs with different degrees of internal subdivision. When an oil tight centerline bulkhead is introduced into a double-hull tanker design, the free surface effect is reduced by a factor of four. That is, the combined free surface of the port and starboard tanks is one-fourth of the free surface of the single tank. The three-tank-across arrangement is typical of many of the small and mid-size single-hull tankers. For a vessel with the proportions shown, the free surface effect is about one-seventh of the "single tank across" arrangement.

For nonrectangular tanks, the free surface effect will vary with the level of the liquid in the tank. For instance, Figure K-7 shows a ballast U tank which is 35 percent full with the water level at one-half the double bottom height, and a tank in which the water level is increased so that the ballast extends into the wings. For the arrangement shown, the free surface effect changes by a factor of three. During cargo handling operations, relatively small changes in ballast can have a dramatic effect on the overall stability.

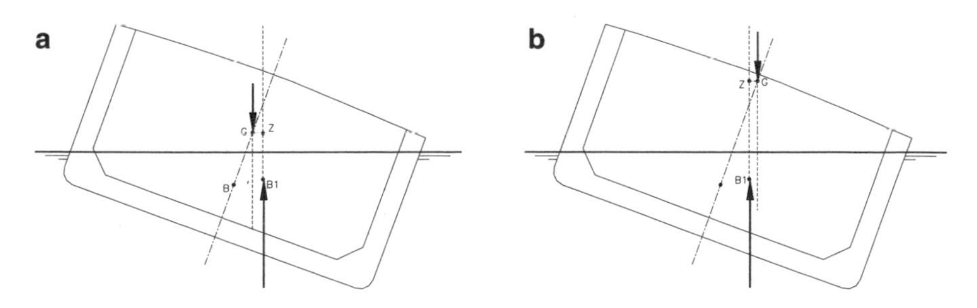

FIGURE K-5 Variation in intact stability. (a) Positive stability. (b) Negative stability.

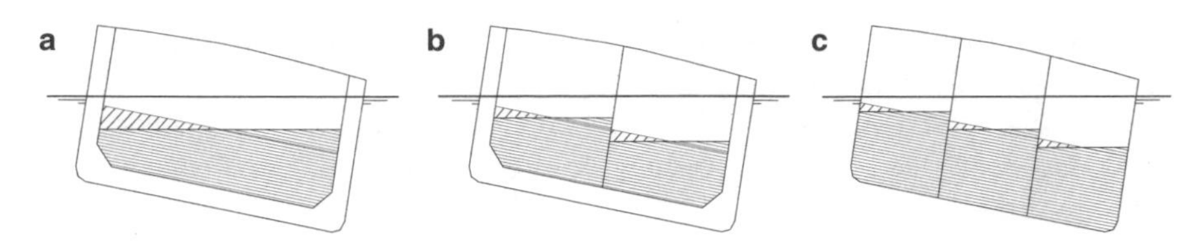

FIGURE K-6 Effect of levels of internal subdivision on free surface effect. (a) Single tank across, free surface effect = 1. (b) With centerline bulkhead, free surface effect = 1/4. (c) Three tanks across, free surface effect = 1/7.

A plot of the GZ values provides a picture of the stability characteristics of a vessel (Figure K-8). The stable vessel shown in Figure K-8 (a) has a positive GZ though 60 degrees heel. The unstable vessel shown in Figure K-8 (b) has a negative GZ and will capsize. Designers and operators often refer to the metacenter height, GM, for an indication of the stability of a vessel in its upright condition. The GM is equal to the slope of the GZ curve at zero degrees heel. The condition shown in Figure K-8 (a) has a positive GM, whereas the condition shown in Figure K-8 (b) has a negative GM.

It is possible for a vessel to be unstable in the upright condition, but attain positive stability as the vessel heels. This phenomenon is illustrated by the GZ plot shown in Figure K-8 (c). The GM is negative and the vessel will tend to heel to one side. It will come to rest at the point when the GZ becomes positive, in this case at 15 degrees. This equilibrium heel angle is referred to as the "angle of loll." If the operator mistakenly assumes that the heel angle is caused by off-center loads rather than negative stability, the operator may decide to add ballast or cargo to the uphill side. The vessel will then abruptly flop to the opposite side, generally assuming an even greater heel angle.

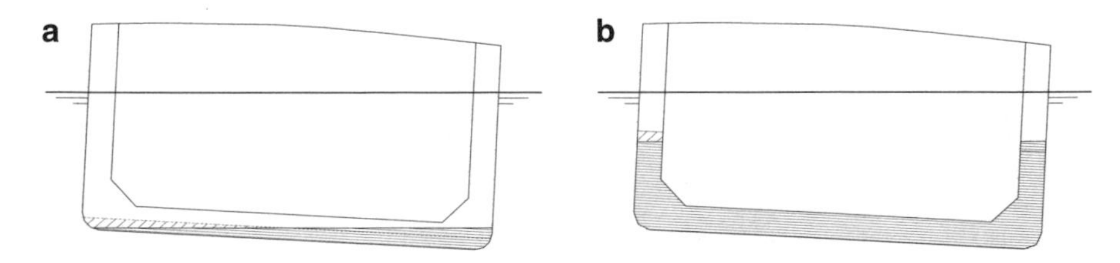

FIGURE K-7 Effect of levels of liquid in tanks on free surface effect. (a) U Tank 35 percent full, free surface effect = 1. (b) U Tank 60 percent full, free surface effect = 1/3.

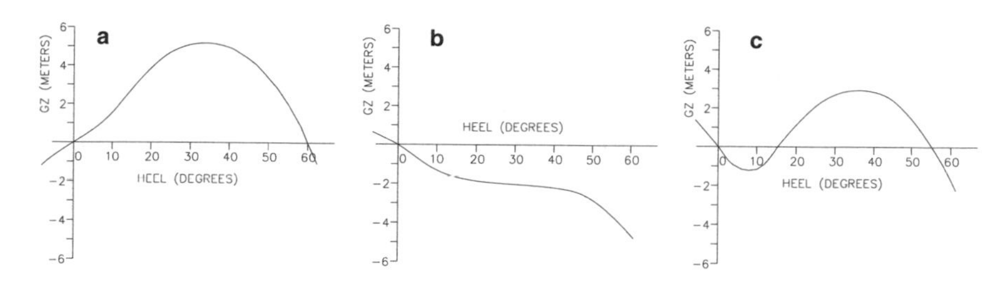

FIGURE K-8 Stability characteristics of a vessel. (a) Vessel with positive stability. (b) Negative stability leading to capsize. (c) Negative stability leading to an angle of loll.

Methodology for Evaluating Intact Stability

For each design, the GM has been calculated for a matrix of load conditions. Uniform loading is assumed at a step size of 1 percent for both cargo tanks and ballast tanks. The free surface correction to GM for the tanks is based upon moment transference for 1 degree heel with 0.9 specific gravity cargo oil and 1.025 specific gravity ballast. Consumable and miscellaneous tanks, such as fuel oil and portable water, are about 50 percent full.

If none of the GM values is less than 0.15 meters, the vessel is assumed to be inherently stable. That is, the vessel will always remain stable regardless of the sequencing of ballast and cargo transfer operations.

If GM values less than 0.15 meters are possible, then the following additional conditions are evaluated:

- The extreme (worst case) load condition stability calculations are performed for the worst case scenario of ballast and cargo loading. Rather than applying free surface effects, liquid transference for each tank is computed at each heel angle. This provides a more accurate assessment of stability at large heel angles. From this calculation, it is determined whether the vessel has any risk of capsize. If the vessel cannot capsize, then the largest possible angle of loll is computed.
- The number of cargo tanks which can be partially full with all ballast tanks at 2 percent filling. The double bottom tanks generally have flat lower surfaces supported by a grillwork of floors and stiffeners, making it difficult to completely strip the tanks of ballast water. Two percent filling has been selected as a readily attainable level of stripping. All ballast tanks are set to 2 percent filling, and all cargo tanks to the level which minimizes GM.
- Even at 2 percent filling, the free surface effects can have a significant impact on stability. If a significant number of cargo tanks must be either empty or 98 percent full in order to maintain positive stability with all ballast tanks at 2 percent filling, then the operating restrictions become

complicated and the risk of operator error increases. Therefore, this condition provides a good indication as to whether satisfactory stability can be maintained through reasonably simple operational restrictions.
- Evaluation of load restrictions to maintain positive stability. A load restriction that would assure positive stability throughout cargo handling operations is developed.

Evaluating Ballast Condition

The international requirements for double-hull tankers are contained in Regulation 13F of the 1992 Amendments to Annex I of MARPOL 73/78 (IMO, 1992). Minimum dimensions for wing tanks and double bottom tanks are specified. This regulation also states that wing tank and double bottom tanks used to meet the IMO ballast draft requirements "shall be located as uniformly as practicable along the cargo tank length."

This requirement tends to produce double-hull tankers with a relatively homogeneous longitudinal distribution of ballast. As compared to most MARPOL 78 tankers where ballast is concentrated closer to amidships, the double-hull tankers can be expected to have higher hogging moments.

In practice, most double-hull tankers are designed with double bottom and wing tank dimensions in excess of the minimum requirements. This is in response to a number of factors: the desire to provide better access into the ballast tanks for inspection and construction purposes, owner requirements to have deeper ballast drafts than the IMO minimum values, and for structural and oil outflow considerations.

Methodology for Evaluating Ballast Condition Longitudinal Strength and Drafts

The fore and aft drafts and the maximum still-water bending moments and shear forces have been computed for the heavy ballast condition. Consumables such as fuel oil and fresh water have been assumed 50 percent full. When allocating ballast, an effort has been made to maximize the forward draft, subject to the following:

- For both the MARPOL tankers and the double-hull tankers, ballast is allocated to segregated ballast tanks only.
- Still-water shear forces and bending moments are maintained within allowable values.
- At least 110 percent propeller immersion is maintained.

The drafts are presented as a percentage of the IMO minimum requirements and as a percentage of propeller immersion. Strength results are presented as a percentage of the allowable values assigned to the vessel by the classification

TABLE K-1 Sizes and Hull Types of Tank Vessels Evaluated

	Single Hull	Double Side	Double Hull
35,000–50,000 DWT tankers	2	1	3
80,000–100,000 DWT tankers	2	1	4
135,000–160,000 DWT tankers	3	—	5
265,000–300,000 DWT tankers	3	—	3
5,000–25,000 DWT barges	5	—	4
Total	15	2	19

society. For comparative purposes, the class assigned permissible still-water bending moment amidships is presented as a percentage of the value obtained by computing a still-water bending moment based on the minimum section modulus, permissible stresses, and assumed wave bending moments contained in Part 3, Section 6 of the American Bureau of Shipping Rules (ABS, 1995). These baseline values are referred to as the ABS standard values in this study.

EVALUATING DESIGN

Table K-1 lists the hull types and numbers of vessels analyzed in this study. Designs have been selected to be representative of the ships and barges trading in U.S. waters. Single-hull tankers in each group include both pre-MARPOL (without segregated ballast tanks) and MARPOL 78 (with segregated ballast tanks in protective locations) vessels. A number of double-side tankers are currently used for lightering services, and therefore a 40,000 DWT and an 85,000 DWT double side tanker have been evaluated. Double-hull tankers in the 35,000 DWT to 160,000 DWT range include vessels with single-tank-across cargo tank arrangements, as well as vessels fitted with tight centerline bulkheads through the cargo block.

Oil outflow, survivability, intact stability, ballast draft, and strength evaluations have been carried out for each tanker. Oil outflow and survivability calculations have also been carried out for each barge.

Evaluating 35,000 DWT–50,000 DWT Tankers

Design Characteristics

Tankers in this size range are often product carriers, with many of the designs having extensive internal subdivision to allow for carriage of a variety of cargoes and grades. Designs above 40,000 DWT generally have a breadth of about 32.2 meters, which is the maximum permitted for normal transit through the Panama Canal. Typical dimensions are as follows:

- Lbp 168.0 m–200.0 m
- beam 27.4 m–32.2 m
- depth 14.8 m–19.1 m
- scantling draft 10.9 m–12.7 m

The cargo blocks for single-hull tankers under 50,000 DWT have historically been arranged three tanks across, and five to eight tanks long. The double-hull tankers built since 1990 have been either one cargo tank across, two cargo tanks across, or a combination thereof. Typical arrangements are shown in Figure K-9.

The single-tank-across designs are usually arranged with seven to nine cargo tanks plus two slop tanks. The two-tanks-across arrangement is generally constructed with twelve (6 × 2) to sixteen (8 × 2) cargo tanks plus two slop tanks.

Three double-hull tanker designs have been evaluated. Design #40-D1 has a single-tank-across arrangement for all cargo oil tanks, and a combination of U and L ballast tanks. Design #40-D2 and #40-D3 are arranged with an oil tight centerline bulkhead fitted over the entire length of the cargo block, and L type ballast tanks. Design #40-D2 has the highest degree of internal subdivision, with an 8 × 2 cargo tank arrangement.

Evaluating 80,000 DWT–100,000 DWT Tankers

Design Characteristics

Typical dimensions for tankers in this size range are as follows:

- Lbp 210.0 m–242.0 m
- beam 38.8 m–44.2 m
- depth 19.2 m–23.2 m
- scantling draft 12.2 m–16.6 m

The cargo blocks for single-hull tankers between 75,000 DWT and 110,000 DWT have been typically arranged three tanks across, and four or five tanks long. Double-side tankers, primarily used as shuttle tankers, generally have single-tank-across cargo tank arrangements, and 4.5 to 6.0 meter-wide wing tanks.

Most of the double-hull vessels between 75,000 DWT and 110,000 DWT are single-tank-across designs, with seven to nine cargo tanks plus two slop tanks. Only a few tankers in this size range have been fitted with oil tight longitudinal bulkheads. Recent designs include arrangements with twelve (6 × 2), fourteen (7 × 2), and eighteen (6 × 3) cargo tanks plus slop tanks.

Four double-hull tanker designs have been evaluated. Design #80-D1 and #80-D2 have single-tank-across arrangements for all cargo oil tanks. Design #80-D3 is a hybrid with a combination of single-tank-across and port and starboard cargo tanks. Design #80-D4 has an oil-tight centerline bulkhead fitted over the entire length of the cargo block. All four designs have L type ballast tanks over the entire length of the cargo block.

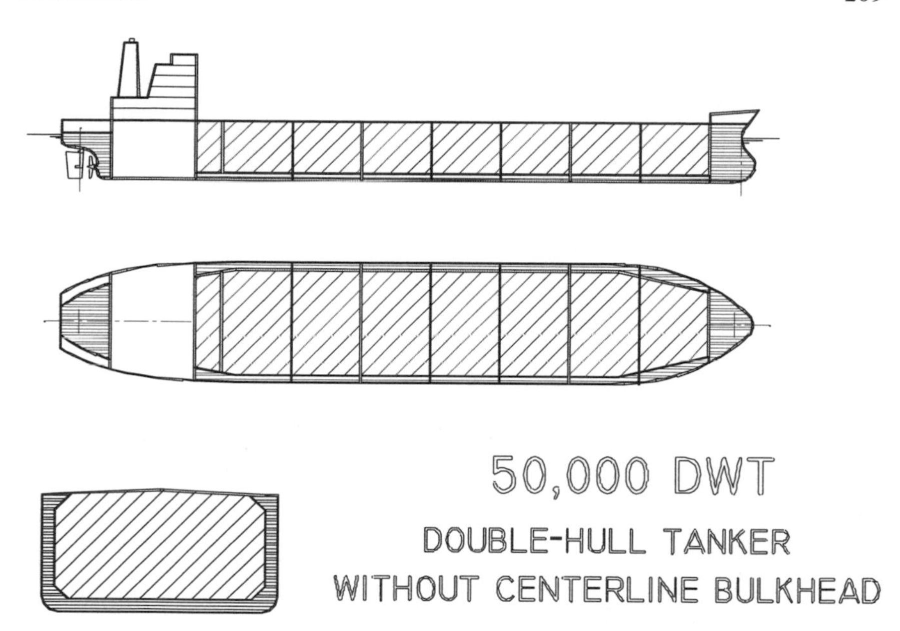

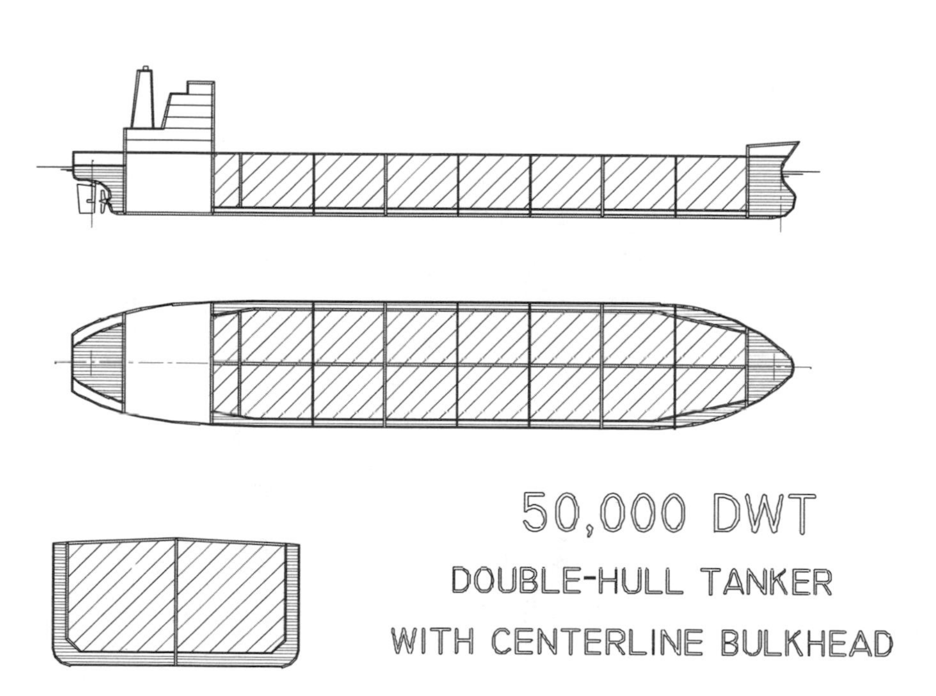

FIGURE K-9 Typical arrangements for 50,000 DWT tanker.

TABLE K-2 Principal Particulars for 35,000 DWT–50,000 DWT Tankers

	#40-S1 pre-MARPOL	#40-S2 MARPOL 78	#40-DS3 Double Sides	#40-D1 Double Hull	#40-D2 Double Hull	#40-D3 Double Hull
Longitudinal Bulkhead in Cargo Tanks	All	All	None	None	All	All
Longitudinal Bulkhead in Ballast Tanks				Some	All	All
Deadweight (MTons)	39,000	36,000	40,000	47,000	40,000	46,000
Length/Beam	7.01	5.53	5.72	5.40	6.51	5.41
Length/Depth	12.92	10.37	12.45	9.67	12.50	9.08
Beam/Depth	1.84	1.88	2.18	1.79	1.92	1.68
Loadline Draft/Depth	0.75	.68	.66	.68	.74	.64
Number of Longitudinal Bulkheads	2	15	6	8	16	14
Number of Cargo Tanks (excl. slops)	13	15	6	8	16	14
Number of Ballast Tanks	7	6	10	9	22	12
Wing Tank Width/Required Width	—	—	2.24	1.13	1.22	1.00
Wing Tank Width/Beam	—	—	0.139	0.070	0.083	0.062
Double Bottom Height/Required Height	—	—	—	1.00	1.09	1.08
Double Bottom Height/Depth	—	—	—	0.111	0.140	0.112
Cargo Oil at 98% (m3)	44,000	44,000	47,000	52,000	43,000	54,000
Segregated Ballast (MTons)	11,000	13,000	20,000	23,000	22,000	20,000

TABLE K-3 Oil Outflow Evaluation for 35,000 DWT–50,000 DWT Tankers

	#40-S1 pre-MARPOL	#40-S2 MARPOL 78	#40-DS3 Double Sides	#40-D1 Double Hull	#40-D2 Double Hull	#40-D3 Double Hull
Side Damage						
Probability of zero outflow	.31	.54	.84	.85	.82	.85
Mean outflow (m^3)	2,180	741	771	1,467	618	803
Extreme (1/10) outflow (m^3)	6,194	4,571	7,509	12,166	4,402	6,739
Combined Bottom Damage [40% 0m: 50% 2m: 10% 6m tide]						
Probability of zero outflow	.13	.10	.12	.83	.84	.84
Mean outflow (m^3)	1,817	2,868	4,156	646	397	560
Extreme (1/10) outflow (m^3)	6,296	8,885	10,406	5,987	3,344	4,914
Combined Side and Bottom Damage [40% Side: 60% Bottom]						
Probability of zero outflow	.20	.28	.41	.84	.83	.84
Mean outflow (m^3)	1,962	2,017	2,802	974	485	657
Extreme (1/10) outflow (m^3)	6,255	7,160	9,247	8,458	3,767	5,644
Pollution Prevention Index						
98% cargo volume (m^3)	41,433	37,929	43,017	50,708	42,764	50,467
Index E	.36	.38	.43	.91	1.06	1.02

TABLE K-4 Survivability Evaluation for 35,000 DWT–50,000 DWT Tankers

	#40-S1 pre-MARPOL	#40-S2 MARPOL 78	#40-DS3 Double Sides	#40-D1 Double Hull	#40-D2 Double Hull	#40-D3 Double Hull
Side Damage Survivability Index	92.5%	97.5%	99.2%	87.2%	100.0%	97.1%

TABLE K-5 Intact Stability Evaluation for 35,000 DWT–50,000 DWT Tankers

	#40-S1 pre-MARPOL	#40-S2 MARPOL 78	#40-DS3 Double Sides	#40-D1 Double Hull	#40-D2 Double Hull	#40-D3 Double Hull
Minimum GMt (m)	1.96	2.93	3.68	–4.03	1.54	0.15
SWB (% capacity)	2%	2%	2%	8%	4%	98%
Cargo oil (% capacity)	98%	98%	98%	9%	97%	4%
Minimum GMt w/ 2% SWB (m)	1.96	2.93	3.68	–0.91	1.54	.96
Possibility of Capsize	none	none	none	none	none	none
Maximum Angle of Loll	none	none	none	15 degrees	none	none
Load restrictions	none	none	none	[a]	none	none

Note: Load Restrictions required to maintain GM greater than 0.15 meters for load/discharge operation:
[a] #40-D1: at least three pair of ballast tanks at 50% or more filling (or)
at least two cargo oil tanks stripped.

TABLE K-6 Ballast Condition Evaluation for 35,000 DWT–50,000 DWT Tankers

	#40-S1 pre-MARPOL	#40-S2 MARPOL 78	#40-DS3 Double Sides	#40-D1 Double Hull	#40-D2 Double Hull	#40-D3 Double Hull
IMO Draft Requirements						
Minimum draft forward (m)	4.4	4.1	4.3	4.2	4.4	4.2
Heavy Ballast Condition						
Number of cargo oil tanks used	3	3	none	none	none	none
Draft aft (m)	8.0	8.0	7.6	8.5	8.9	7.5
Propeller immersion	117%	141%	124%	138%	143%	121%
Draft forward (m)	5.0	4.5	4.4	6.3	6.6	6.9
Draft forward as % of IMO required	113%	110%	102%	151%	150%	165%
Allowable Bending Moment (Hog)						
As % of ABS minimum value	98%	123%	97%	100%	118%	153%
Heavy Ballast Condition						
Maximum bending moment	39%	75%	54%	98%	96%	56%
Maximum shear force	43%	30%	24%	82%	40%	11%

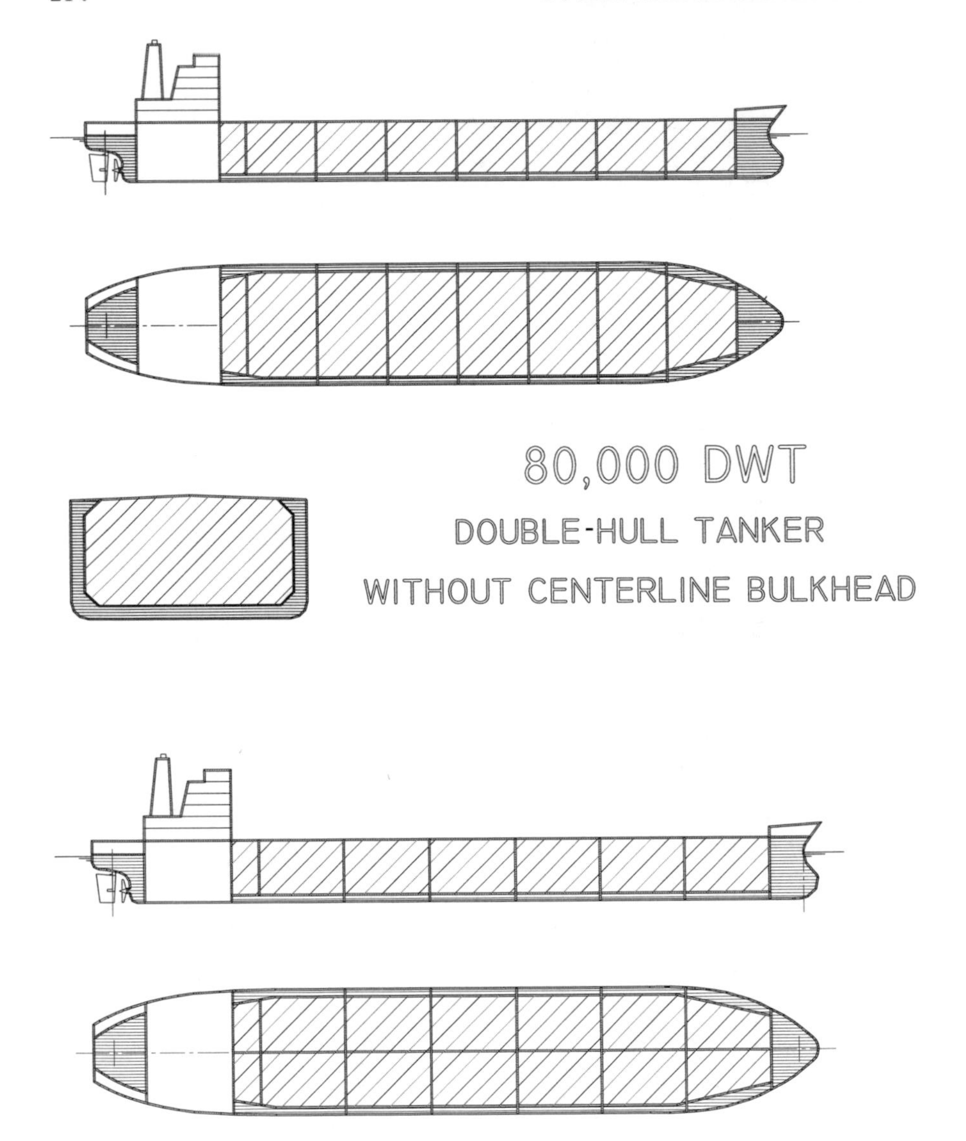

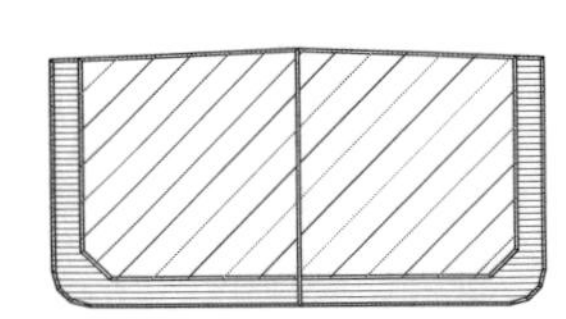

FIGURE K-10 Typical arrangements for 80,000 DWT tankers.

TABLE K-7 Principal Particulars for 80,000 DWT–100,000 DWT Tankers

	#80-S1 pre-MARPOL	#80-S2 MARPOL 78	#80-DS3 Double Sides	#80-D1 Double Hull	#80-D2 Double Hull	#80-D3 Double Hull	#80-D4 Double Hull
Longitudinal Bulkhead in Cargo Tanks	All	All	None	None	None	Some	All
Longitudinal Bulkhead in Ballast Tanks				All	All	All	All
Deadweight (MTons)	97,000	81,000	85,000	97,000	94,000	96,000	96,000
Length/Beam	5.75	5.93	5.09	5.57	5.57	5.60	5.57
Length/Depth	11.62	12.46	12.38	11.65	12.00	12.05	11.65
Beam/Depth	2.02	2.10	2.43	2.09	2.15	2.15	2.09
Loadline Draft/Depth	0.75	0.64	0.67	0.68	0.70	0.69	0.69
Number of Longitudinal Bulkheads	2	2	2	2	2	3	3
Number of Cargo Tanks (excl. slops)	12	11	7	7	7	10	12
Number of Ballast Tanks	4	8	10	10	12	13	14
Wing Tank Width/Required Width	—	—	3.00	1.00	1.35	1.45	1.23
Wing Tank Width/Beam	—	—	0.136	0.048	0.064	0.069	0.059
Double Bottom Height/Required Height	—	—	—	1.05	1.05	1.05	1.23
Double Bottom Height/Depth	—	—	—	0.105	0.108	0.108	0.123
Cargo Oil at 98% (m^3)	119,000	100,000	98,000	108,000	106,000	108,000	108,000
Segregated Ballast (MTons)	13,000	38,000	40,000	41,000	40,000	39,000	42,000

TABLE K-8 Oil Outflow Evaluation for 80,000 DWT–100,000 DWT Tankers

	#80-S1 pre-MARPOL	#80-S2 MARPOL 78	#80-DS3 Double Sides	#80-D1 Double Hull	#80-D2 Double Hull	#80-D3 Double Hull	#80-D4 Double Hull
Side Damage							
Probability of zero outflow	.22	.32	.92	.81	.85	.87	.83
Mean outflow (m^3)	9,508	6,180	1,411	4,369	3,170	2,253	2,013
Extreme (1/10) outflow (m^3)	20,750	13,311	14,105	30,634	25,586	19,988	15,071
Combined Bottom Damage [40% 0m: 50% 2m: 10% 6m tide]							
Probability of zero outflow	.09	.09	.11	.82	.81	.80	.82
Mean outflow (m^3)	4,564	6,814	8,939	1,485	1,706	1,497	1,093
Extreme (1/10) outflow (m^3)	13,813	17,097	24,155	12,726	13,348	11,524	8,828
Combined Side and Bottom Damage [40% Side: 60% Bottom]							
Probability of zero outflow	.14	.19	.44	.82	.82	.83	.83
Mean outflow (m^3)	6,542	6,560	5,928	2,639	2,292	1,799	1,461
Extreme (1/10) outflow (m^3)	16,588	15,583	20,135	19,889	18,243	14,910	11,325
Pollution Prevention Index							
98% cargo volume (m^3)	104,749	81,227	92,276	105,826	102,081	104,169	105,269
Index E	.29	.28	.46	.85	.88	.96	1.03

TABLE K-9 Survivability Evaluation for 80,000 DWT–100,000 DWT Tankers

	#80-S1 pre-MARPOL	#80-S2 MARPOL 78	#80-DS3 Double Sides	#80-D1 Double Hull	#80-D2 Double Hull	#80-D3 Double Hull	#80-D4 Double Hull
Side Damage Survivability Index	99.2%	100.0%	100.0%	99.2%	99.7%	99.9%	99.9%

TABLE K-10 Intact Stability Evaluation for 80,000 DWT–100,000 DWT Tankers

	#80-S1 pre-MARPOL	#80-S2 MARPOL 78	#80-DS3 Double Sides	#80-D1 Double Hull	#80-D2 Double Hull	#80-D3 Double Hull	#80-D4 Double Hull
Minimum GMt (m)	5.27	5.66	8.77	–2.79	0.41	0.15	2.98
SWB (% capacity)	2%	2%	2%	4%	4%	5%	4%
Cargo oil (% capacity)	97%	98%	98%	72%	90%	91%	98%
Minimum GMt w/ 2% SWB (m)	5.27	5.66	8.77	0.49	1.05	3.03	3.54
Possibility of Capsize	none	none	none	none	none	none	none
Maximum Angle of Loll	none	none	none	3 degrees	none	none	none
Load restrictions	none	none	none	[a]	none	none	none

Note: Load Restrictions required to maintain GM greater than 0.15 meters for load/discharge operation:
[a] #80-D1: at least one pair of ballast tanks at 50% or more filling.

TABLE K-11 Ballast Condition Evaluation for 80,000 DWT–100,000 DWT Tankers

	#80-S1 pre-MARPOL	#80-S2 MARPOL 78	#80-DS3 Double Sides	#80-D1 Double Hull	#80-D2 Double Hull	#80-D3 Double Hull	#80-D4 Double Hull
IMO Draft Requirements							
Minimum draft forward (m)	4.9	5.0	4.8	4.9	4.9	4.9	4.9
Heavy Ballast Condition							
Number of cargo oil tanks used	3	3	none	none	none	none	none
Draft aft (m)	9.4	8.7	8.3	8.3	9.1	8.9	8.9
Propeller immersion	124%	114%	112%	117%	113%	113%	127%
Draft forward (m)	6.3	5.7	6.3	6.1	6.0	5.9	6.0
Draft forward as % of IMO required	129%	115%	131%	124%	122%	119%	123%
Allowable Bending Moment (Hog)							
As % of ABS minimum value	106%	91%	96%	91%	98%	120%	91%
Heavy Ballast Condition							
Maximum bending moment	37%	77%	84%	100%	100%	86%	100%
Maximum shear force	3%	61%	64%	67%	67%	35%	62%

Evaluating 135,000 DWT–160,000 DWT Tankers

Design Characteristics

Tankers in this size range are generally designed to the maximum proportions suitable for passage through the Suez Canal. Typical dimensions are as follows:

- lbp 258.0 m–265.0 m
- beam 43.0 m–50.0 m
- depth 22.8 m–25.8 m
- scantling draft 15.2 m–17.2 m

The cargo blocks for single-hull SUEZMAX tankers have historically been arranged three tanks across, and five or six tanks long. The double-hull SUEZMAX tankers built since 1990 have been either one cargo tank across, two cargo tanks across, or a combination thereof. Typical arrangements are shown in Figure K-11.

The single-tank-across designs are usually arranged with nine cargo tanks plus two slop tanks, which is the maximum tank size meeting the IMO tank size and outflow requirements as defined in Regulations 22-24 of Annex I to MARPOL 73/78.

The two-tanks-across arrangement is generally constructed with ten (5 × 2) or twelve (6 × 2) cargo tanks plus two slop tanks. Although the 5 × 2 arrangement satisfies IMO requirements, damage stability requirements impose some operating restrictions with regard to deep draft conditions with partially full cargo tanks. This, together with considerations for greater segregation of cargoes, has led many ship owners to opt for the 6 × 2 cargo tank arrangement.

Five double-hull tanker designs have been evaluated. Design #150-D1 has a single-tank-across arrangement for all cargo oil tanks. Design #150-D2 is a hybrid, with four single-tank-across cargo tanks and three pairs of port and starboard cargo tanks. Designs #150-D3 through #150-D5 all have an oil-tight centerline bulkhead fitted over the entire length of the cargo block. Design #150-D5 has relatively wide wing tanks and a deep double bottom. In order for design #150-D5 to meet the IMO two compartment and raking bottom damage stability requirements, approximately 60 percent of the ballast capacity within the cargo block length is arranged in U tanks.

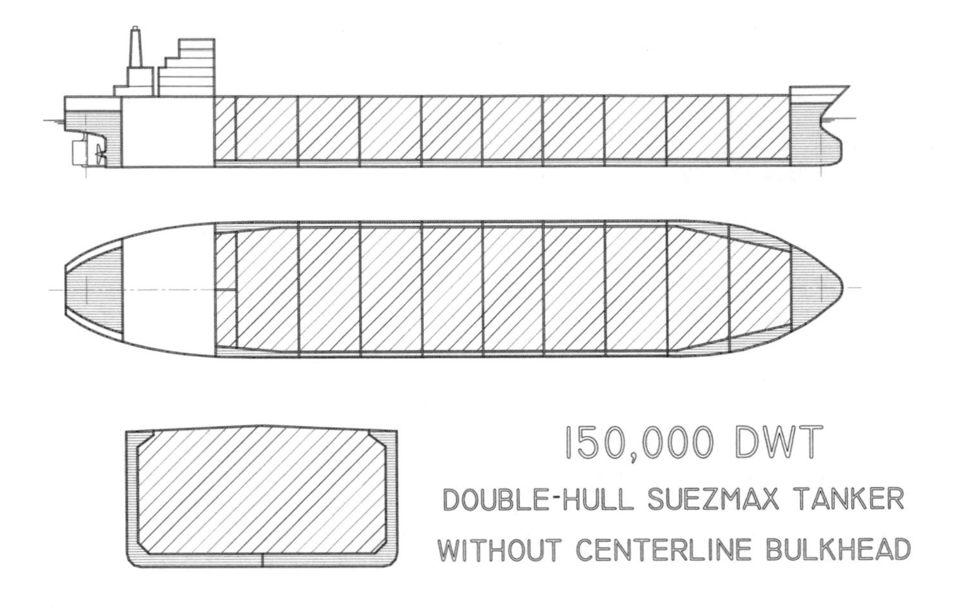

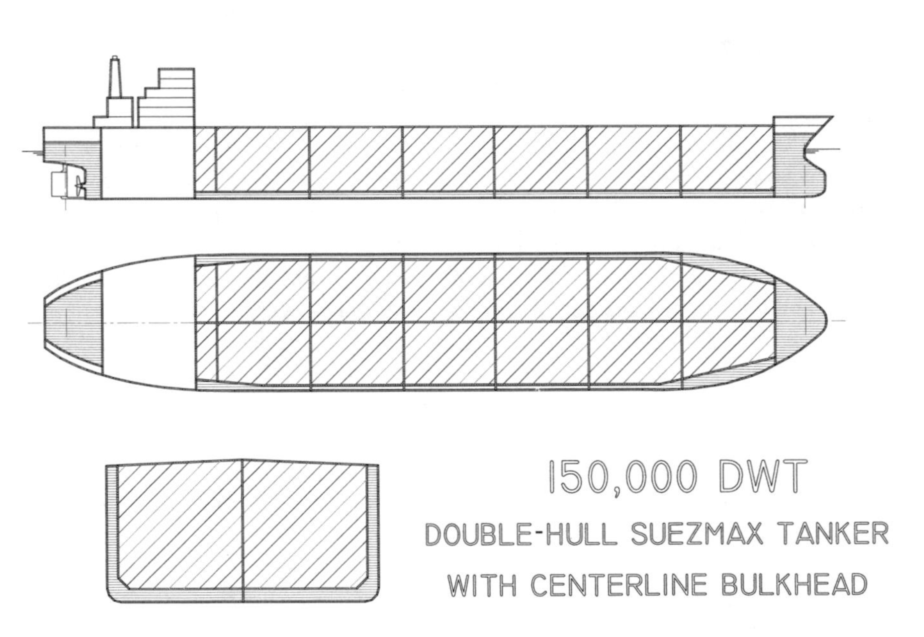

FIGURE K-11 Typical arrangements for 150,000 DWT tankers.

TABLE K-12 Principal Particulars for 135,000 DWT–160,000 DWT Tankers

	#150-S1 pre-MARPOL	#150-S2 MARPOL 78	#150-S3 MARPOL 78	#150-D1 Double Hull	#150-D2 Double Hull	#150-D3 Double Hull	#150-D4 Double Hull	#150-D5 Double Hull
Longitudinal Bulkhead in Cargo Tanks	All	All	All	None	Some	All	All	All
Longitudinal Bulkhead in Ballast Tanks				All	Some	All	All	Some
Deadweight (MTons)	156,000	149,000	155,000	151,000	136,000	150,000	150,000	157,000
Length/Beam	5.00	5.78	5.22	5.50	5.08	5.61	5.52	5.22
Length/Depth	13.40	11.81	10.40	11.48	9.75	10.79	11.58	10.40
Beam/Depth	2.68	2.04	1.99	2.09	1.92	1.92	2.10	1.99
Loadline Draft/Depth	0.77	0.66	0.67	0.70	0.67	0.71	0.70	0.69
Number of Longitudinal Bulkheads	2	2	2	2	3	3	3	3
Number of Cargo Tanks (excl. slops)	12	9	11	9	10	12	12	12
Number of Ballast Tanks	4	8	6	12	16	14	14	15
Wing Tank Width/Required Width	—	—	—	1.15	1.88	1.35	1.28	1.67
Wing Tank Width/Beam	—	—	—	0.048	0.078	0.059	0.053	0.067
Double Bottom Height/Required Height	—	—	—	1.25	1.65	1.40	1.40	1.67
Double Bottom Height/Depth	—	—	—	0.109	0.131	0.117	0.123	0.133
Cargo Oil at 98% (m^3)	187,000	177,000	180,000	165,000	159,000	164,000	167,000	179,000
Segregated Ballast (MTons)	24,000	64,000	66,000	59,000	63,000	56,000	57,000	67,000

TABLE K-13 Oil Outflow Evaluation for 135,000 DWT–160,000 DWT Tankers

	#150-S1 pre-MARPOL	#150-S2 MARPOL 78	#150-S3 MARPOL 78	#150-D1 Double Hull	#150-D2 Double Hull	#150-D3 Double Hull	#150-D4 Double Hull	#150-D5 Double Hull
Side Damage								
Probability of zero outflow	.24	.40	.34	.79	.87	.82	.80	.81
Mean outflow (m^3)	10,868	9,175	8,404	6,436	3,674	3,182	3,481	3,252
Extreme (1/10) outflow (m^3)	27,712	20,406	18,161	43,106	30,789	23,699	23,837	23,524
Combined Bottom Damage [40% 0m: 50% 2m: 10% 6m tide]								
Probability of zero outflow	.08	.08	.08	.80	.83	.82	.82	.83
Mean outflow (m^3)	7,965	11,600	11,725	2,652	2,009	1,663	1,605	1,520
Extreme (1/10) outflow (m^3)	20,945	29,795	29,367	21,183	16,049	13,060	12,638	12,623
Combined Side and Bottom Damage [40% Side: 60% Bottom]								
Probability of zero outflow	.14	.21	.18	.80	.84	.82	.81	.82
Mean outflow (m^3)	9,126	10,630	10,396	4,166	2,675	2,271	2,355	2,213
Extreme (1/10) outflow (m^3)	23,652	26,039	24,885	29,952	21,945	17,316	17,118	16,983
Pollution Prevention Index								
98% cargo volume (m^3)	186,692	162,063	168,926	164,895	148,206	163,619	162,225	170,961
Index E	.36	.34	.34	.87	.99	1.06	1.04	1.09

TABLE K-14 Survivability Evaluation for 135,000 DWT–160,000 DWT Tankers

	#150-S1 pre-MARPOL	#150-S2 MARPOL 78	#150-S3 MARPOL 78	#150-D1 Double Hull	#150-D2 Double Hull	#150-D3 Double Hull	#150-D4 Double Hull	#150-D4 Double Hull
Side Damage Survivability Index	99.9%	100.0%	100.0%	99.8%	99.2%	99.9%	100.0%	99.5%

TABLE K-15 Intact Stability Evaluation for 135,000 DWT–160,000 DWT Tankers

	#150-S1 pre-MARPOL	#150-S2 MARPOL 78	#150-S3 MARPOL 78	#150-D1 Double Hull	#150-D2 Double Hull	#150-D3 Double Hull	#150-D4 Double Hull	#150-D5 Double Hull
Minimum GMt (m)	11.7	6.37	6.70	–0.53	–2.51	2.74	3.86	0.77
SWB (% capacity)	24%	2%	2%	5%	26%	5%	5%	14%
Cargo oil (% capacity)	88%	98%	98%	72%	10%	98%	97%	97%
Minimum GMt w/ 2% SWB (m)	11.99	6.37	6.70	0.42	0.69	3.50	4.62	2.92
Possibility of Capsize	none	none	none	none	none	none	none	none
Maximum Angle of Loll	none	none	none	5 degrees	8 degrees	none	none	none
Load restrictions	none	none	none	[a]	[b]	none	none	none

Note: Load Restrictions required to maintain GM greater than 0.15 meters for load/discharge operation:

[a] #150-D1: at least one pair of L tanks at 50% or more filling

[b] #150-D2: at least two U ballast tanks at 50% or more filling.

TABLE K-16 Ballast Condition Evaluation for 135,000 DWT–160,000 DWT Tankers

	#150-S1 pre-MARPOL	#150-S2 MARPOL 78	#150-S3 MARPOL 78	#150-D1 Double Hull	#150-D2 Double Hull	#150-D3 Double Hull	#150-D4 Double Hull	#150-D5 Double Hull
IMO Draft Requirements								
Minimum draft forward (m)	5.4	5.5	5.3	5.3	5.1	5.2	5.3	5.3
Heavy Ballast Condition								
Number of cargo oil tanks used	3	none	none	none	none	none	none	none
Draft sft (m)	11.5	10.3	10.2	9.0	11.0	9.9	7.9	10.9
Propeller immersion	147%	126%	126%	110%	139%	115%	106%	135%
Draft forward (m)	6.9	6.9	8.3	6.9	9.2	7.1	7.3	8.2
Draft forward as % of IMO required	130%	126%	158%	130%	181%	135%	138%	156%
Allowable Bending Moment (Hog)								
As % of ABS minimum value	94%	89%	83%	102%	109%	127%	94%	99%
Heavy Ballast Condition								
Maximum bending moment	85%	100%	56%	100%	100%	100%	100%	100%
Maximum shear force	60%	89%	69%	53%	56%	90%	48%	53%

Evaluating 265,000 DWT–300,000 DWT Tankers

Design Characteristics

Typical dimensions for VLCCs are as follows:

- Lbp 315.0 m–326.0 m
- beam 53.0 m–68.0 m
- depth 26.0 m–32.0 m
- scantling draft 19.0 m–23.0 m

A majority of the single-hull VLCCs have a 5 long × 3 wide cargo tank arrangement. The pre-MARPOL designs typically have one or two ballast tanks within the cargo block, whereas the MARPOL 78 designs usually have wing ballast tanks port and starboard at the No.2 and No.4 positions. Variations include a few tankers with 4 × 3 cargo tank arrangements at the lower end of the size range, and some vessels with 6 × 3 cargo tank arrangements.

Most of the double-hull designs built since 1990 are arranged with 5 × 3 cargo tanks plus slop tanks. The double bottom depth is typically about 3 meters, and the wing tank widths vary from 3 to 4 meters. A typical arrangement is shown in Figure K-12.

Three double-hull tanker designs have been evaluated. All three have a 5 × 3 cargo tank arrangement. Design #280-D1 has all L ballast tanks, design #280-D2 has predominantly L ballast tanks with one U tank. Design #280-D3 has predominantly full-breadth double bottom ballast tanks with independent side tanks port and starboard, together with midship ballast tanks arranged inboard of the longitudinal bulkheads.

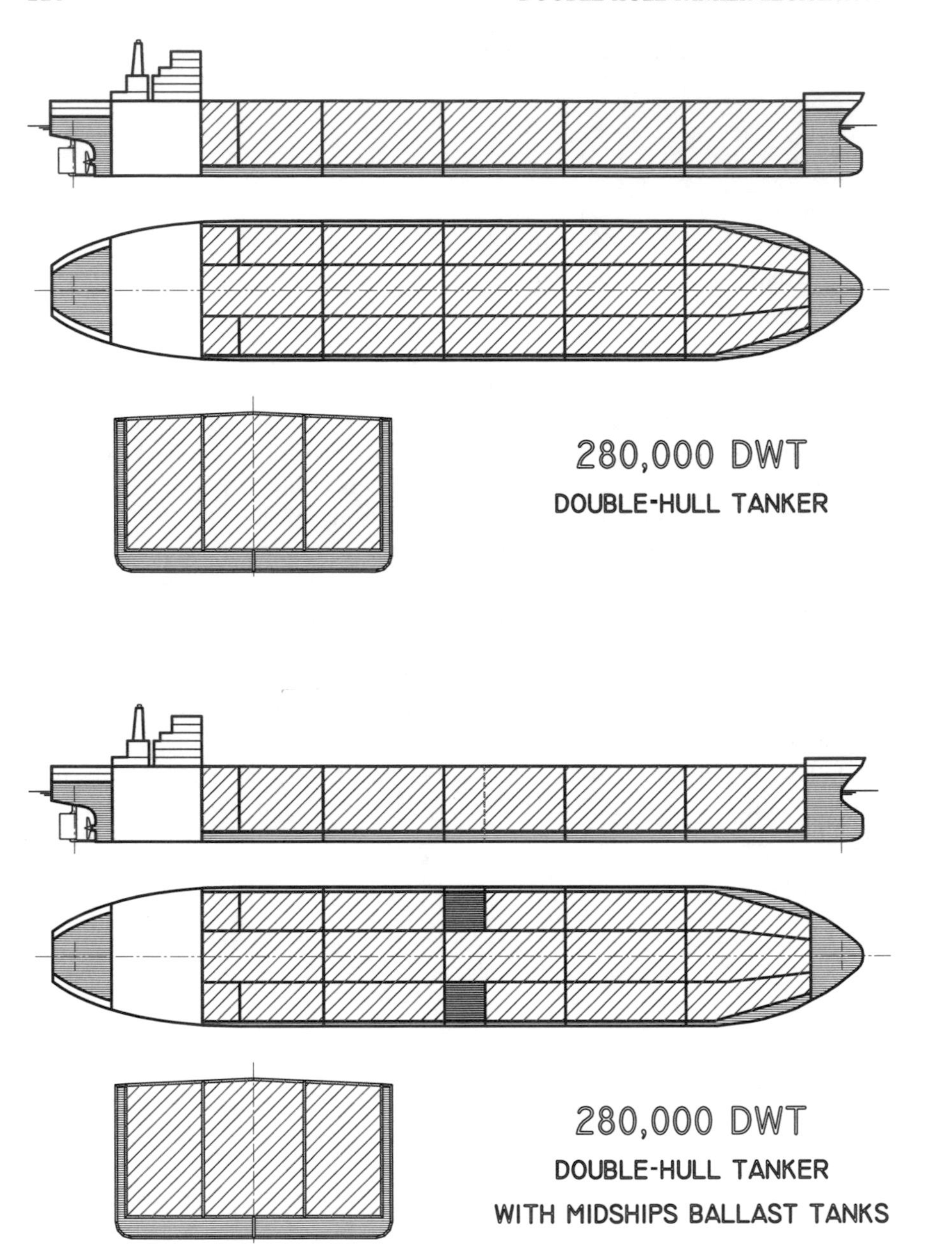

FIGURE K-12 Typical arrangements for 280,000 DWT tankers.

TABLE K-17 Principal Particulars for 265,000 DWT–300,000 DWT Tankers

	#280-S1 pre-MARPOL	#280-S2 MARPOL 78	#280-S3 MARPOL 78	#280-D1 Double Hull	#280-D2 Double Hull	#280-D3 Double Hull
Longitudinal Bulkhead in Cargo Tanks	All	All	All	All	All	All
Longitudinal Bulkhead in Ballast Tanks				All	Some	None
Deadweight (MTons)	277,000	268,000	292,000	280,000	300,000	298,000
Length/Beam	5.87	5.97	5.50	5.47	5.52	5.37
Length/Depth	12.31	12.12	10.13	10.10	10.32	10.06
Beam/Depth	2.10	2.03	1.84	1.85	1.87	1.87
Loadline Draft/Depth	0.81	0.78	0.69	0.66	0.71	0.70
Number of Longitudinal Bulkheads	2	2	2	4	4	4
Number of Cargo Tanks (excl. slops)	16	12	13	15	15	15
Number of Ballast Tanks	3	6	6	13	14	17
Wing Tank Width/Required Width	—	—	—	1.97	3.52	3.15
Wing Tank Width/Beam	—	—	—	0.068	0.061	0.053
Double Bottom Height/Required Height	—	—	—	1.55	1.50	1.60
Double Bottom Height/Depth	—	—	—	0.099	0.097	0.101
Cargo Oil at 98% (m^3)	309,000	314,000	338,000	343,000	338,000	244,000
Segregated Ballast (MTons)	37,000	41,000	114,000	108,000	102,000	111,000

TABLE K-18 Oil Outflow Evaluation for 265,000 DWT–300,000 DWT Tankers

	#280-S1 pre-MARPOL	#280-S2 MARPOL 78	#280-S3 MARPOL 78	#280-D1 Double Hull	#280-D2 Double Hull	#280-D3 Double Hull
Side Damage						
Probability of zero outflow	.15	.19	.33	.82	.81	.72
Mean outflow (m^3)	23,448	21,617	15,342	4,778	4,474	6,134
Extreme (1/10) outflow (m^3)	56,162	49,224	35,107	34,226	31,393	35,325
Combined Bottom Damage [40% 0m: 50% 2m: 10% 6m tide]						
Probability of zero outflow	.08	.07	.07	.80	.79	.81
Mean outflow (m^3)	9,392	10,668	15,956	2,949	2,420	2,704
Extreme (1/10) outflow (m^3)	26,254	27,190	50,624	22,731	19,283	23,039
Combined Side and Bottom Damage [40% Side: 60% Bottom]						
Probability of zero outflow	.11	.12	.17	.81	.80	.78
Mean outflow (m^3)	15,014	15,047	15,710	3,681	3,242	4,076
Extreme (1/10) outflow (m^3)	38,217	36,003	44,417	27,329	24,127	27,953
Pollution Prevention Index						
98% cargo volume (m^3)	299,561	267,062	319,218	306,572	328,112	325,272
Index E	.30	.28	.33	1.04	1.09	1.01

TABLE K-19 Survivability Evaluation for 265,000 DWT–300,000 DWT Tankers

	#280-S1 pre-MARPOL	#280-S2 MARPOL 78	#280-S3 MARPOL 78	#280-D1 Double Hull	#280-D2 Double Hull	#280-D3 Double Hull
Side Damage Survivability Index	100.0%	100.0%	99.7%	100.0%	100.0%	100.0%

TABLE K-20 Intact Stability Evaluation for 265,000 DWT–300,000 DWT Tankers

	#280-S1 pre-MARPOL	#280-S2 MARPOL 78	#280-S3 MARPOL 78	#280-D1 Double Hull	#280-D2 Double Hull	#280-D3 Double Hull
Minimum GMt (m)	6.94	7.40	6.76	2.40	4.08	0.15
SWB (% capacity)	2%	2%	2%	9%	5%	39%
Cargo oil (% capacity)	80%	97%	98%	98%	98%	98%
Minimum GMt w/ 2% SWB (m)	6.94	7.4	6.76	4.05	4.84	5.30
Possibility of Capsize	none	none	none	none	none	none
Maximum Angle of Loll	none	none	none	none	none	none
Load restrictions	none	none	none	none	none	none

TABLE K-21 Ballast Condition Evaluation for 265,000 DWT–300,000 DWT Tankers

	#280-S1 pre-MARPOL	#280-S2 MARPOL 78	#280-S3 MARPOL 78	#280-D1 Double Hull	#280-D2 Double Hull	#280-D3 Double Hull
IMO Draft Requirements						
Minimum draft forward (m)	6.0	6.0	6.0	6.0	6.0	6.0
Heavy Ballast Condition						
Number of cargo oil tanks used	3	3	none	none	none	none
Draft aft (m)	11.6	10.7	12.2	12.5	12.9	13.1
Propeller immersion	110%	104%	111%	119%	129%	126%
Draft forward (m)	8.6	8.4	9.6	8.4	7.9	9.4
Draft forward as % of IMO required	144%	140%	161%	141%	132%	158%
Allowable Bending Moment (Hog)						
As % of ABS minimum value	86%	87%	106%	105%	106%	119%
Heavy Ballast Condition						
Maximum bending moment	70%	90%	59%	100%	100%	90%
Maximum shear force	64%	85%	100%	53%	79%	57%

Oceangoing Barges

Design Characteristics

Oceangoing barges operating in U.S. waters tend to be smaller than tankers, with few barges exceeding 25,000 DWT. Barges are subject to less stringent loadline requirements than self-propelled tank ships, and will generally have a lower freeboard. When barges are carrying lighter crudes and products, it is not usual for the cargo oil to be in hydrostatic balance relative to the sea.

Single-hull barges above 5,000 DWT are generally arranged with one and sometimes two longitudinal bulkheads. The cargo tank arrangement will vary depending on the extent of cargo segregation required. Common arrangements include (4 × 2) up to (8 × 2) cargo tanks, with a few vessels featuring three-wide cargo tank configurations.

Barges may be of the flush deck type, or fitted with a raised trunk as shown in Figure K-13. Voids are arranged fore and aft within the rake. Oceangoing barges are generally pushed or pulled by tugs, and are often constructed with a notch aft. Ballast tanks may be of the L or U type. They are generally left as void spaces.

Four double-hull tank barges have been evaluated. Designs #B35-D1, B90-D1, and B90-D2 are new barges constructed in the last five years. Design #B179-D1 is a proposed conversion of #B179-S1, an existing 23,700 DWT single-hull barge.

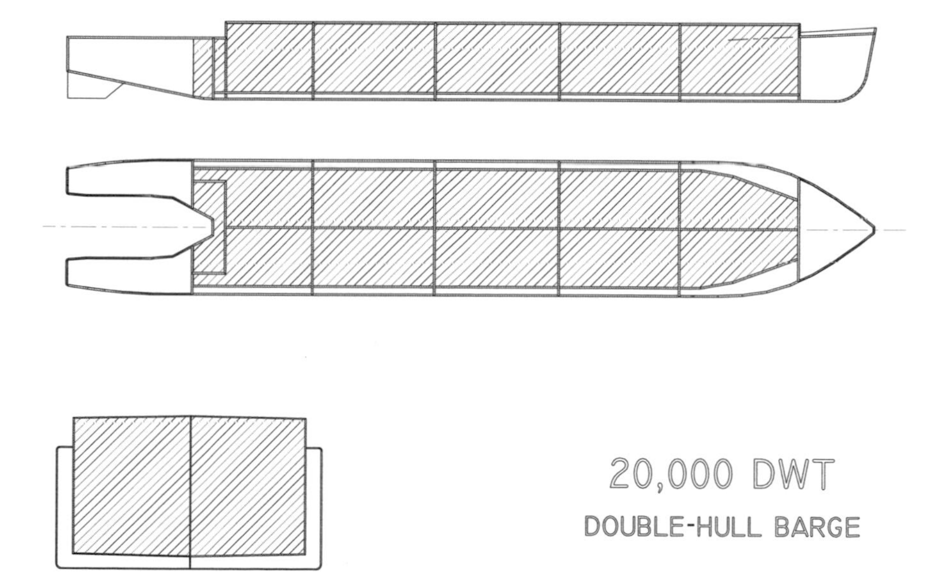

FIGURE K-13 Typical arrangement for double-hull oceangoing barges.

TABLE K-22 Principal Particulars for Oceangoing Barges

	#B35-S1 Single Hull	#B35-D1 Double Hull	#B90-S1 Single Hull	#B90-S2 Single Hull	#B90-S3 Single Hull	#B90-D1 Double Hull	#B90-D2 Double Hull	#B170-S1 Single Hull	#B170-D1 Double Hull
Longitudinal Bulkhead in Cargo Tanks	All	All	All	All	All	All	All	All	All
Longitudinal Bulkhead in Ballast Tanks		None				All	All		All
Deadweight (MTons)	5,500	5,100	11,100	11,100	12,900	11,500	12,800	23,700	22,800
Type	Flush Dk	Flush Dk	Flush Dk	Flush Dk	Flush Dk	with Trunk	Flush Dk	Flush Dk	with Trunk
Length/Beam	3.29	5.55	3.36	3.36	4.86	13.28	5.40	13.70	13.70
Length/Depth	15.00	17.03	13.76	13.76	12.25	2.96	13.03	2.40	2.4
Beam/Depth	4.56	3.07	4.09	4.09	2.52	0.85	2.41	0.86	0.86
Loadline Draft/Depth	0.81	0.85	0.84	0.84	0.88	4.49	0.68	5.71	5.71
Number of Longitudinal Bulkheads	1	3	2	1	1	3	3	1	3
Number of Cargo Tanks (excl. slops)	4 × 2	4 × 2	3 × 3	4 × 2	7 × 2	5 × 2	7 × 2	6 × 2	6 × 2
Number of Ballast Tanks	0	0	1	1	0	0	0	0	13
Wing Tank Width/Required Width	—	1.22	—	—	—	1.13	1.32	—	1.24
Wing Tank Width/Beam	—	0.074	—	—	—	0.054	0.066	—	0.079
Double Bottom Height/Required Height	—	0.694	—	—	—	1.062	1.047	—	1.081
Double Bottom Height/Depth	—	0.039	—	—	—	0.042	0.034	—	0.031
Cargo Oil at 98% (m^3)	6,080	5,690	12,320	12,320	16,040	12,980	14,210	28,520	26,570
Segregated Ballast (MTons)	38,242	35,789	77,490	77,490	100,888	81,642	89,378	179,385	167,120

TABLE K-23 Oil Outflow Evaluation for Oceangoing Barges

	#B35-S1 Single Hull	#B35-D1 Double Hull	#B90-S1 Single Hull	#B90-S2 Single Hull	#B90-S3 Single Hull	#B90-D1 Double Hull	#B90-D2 Double Hull	#B170-S1 Single Hull	#B170-D1 Double Hull
Side Damage									
Probability of zero outflow	.24	.87	.19	.19	.03	.80	.85	.12	.87
Mean outflow (m^3)	727	113	1,024	1,558	1,580	334	229	2,666	352
Extreme (1/10) outflow (m^3)	1,566	921	2,409	3,153	3,048	2,100	1,830	5,307	3,215
Combined Bottom Damage [40% 0m: 50% 2m: 10% 6m tide]									
Probability of zero outflow	.23	.78	.11	.11	.03	.90	.87	.05	.87
Mean outflow (m^3)	581	135	1.046	1,040	648	140	152	1,249	284
Extreme (1/10) outflow (m^3)	1,697	956	2.662	2,702	1,679	1,388	1,394	3,244	2,611
Combined Side and Bottom Damage [40% Side: 60% Bottom]									
Probability of zero outflow	.24	.81	.14	.14	.03	.86	.86	.08	.87
Mean outflow (m^3)	639	126	1,037	1,247	1,021	217	183	1,816	311
Extreme (1/10) outflow (m^3)	1,645	942	2,561	2,882	2,226	1,673	1,568	4,069	2,852
Pollution Prevention Index									
98% cargo volume (m^3)	11,627	10,888	23,568	23,568	27,463	24,809	27,171	50,297	48,425
Index E	.40	1.13	.39	.35	.37	1.25	1.33	.39	1.31

TABLE K-24 Survivability Evaluation for Oceangoing Barges

	#B35-S1 Single Hull	#B35-D1 Double Hull	#B90-S1 Single Hull	#B90-S2 Single Hull	#B90-S3 Single Hull	#B90-D1 Double Hull	#B90-D2 Double Hull	#B170-S1 Single Hull	#B170-D1 Double Hull
Side Damage Survivability Index	95.0%	96.7%	92.9%	92.9%	99.7%	99.5%	99.9%	99.0%	95.0%

SUMMARY AND OBSERVATIONS

Observations on Oil Outflow Analysis of Tankers

The probability of zero outflow is a measure of a tanker's ability to avoid oil spills. In this regard, double-hull tankers perform significantly better than single-hull tankers, as the protective double skin reduces the number of casualties that penetrate into the cargo tanks. As shown in Figure K-14, the probability of zero outflow is four to six times higher for double-hull tankers, indicating single-hull tankers involved in a collision or grounding will be four to six times more likely to spill oil.

The probability of zero outflow is a function of the double bottom and wing tank dimensions, and is not affected by the internal subdivision within the cargo tanks. Therefore, centerline or other longitudinal bulkheads within the cargo spaces have no influence on the probability of zero outflow.

The mean outflow is a measure of the ability of a design to mitigate the amount of oil outflow. Again, double hulls perform significantly better than single-hull vessels, with double-hull mean outflow values averaging one-third to one-fourth of the single-hull values.

The double-side vessels (#40-DS3 and #80-DS3) perform reasonably well with respect to collisions, but have higher outflows for bottom damage. These vessels have single-tank-across arrangements for cargo tanks, which significantly

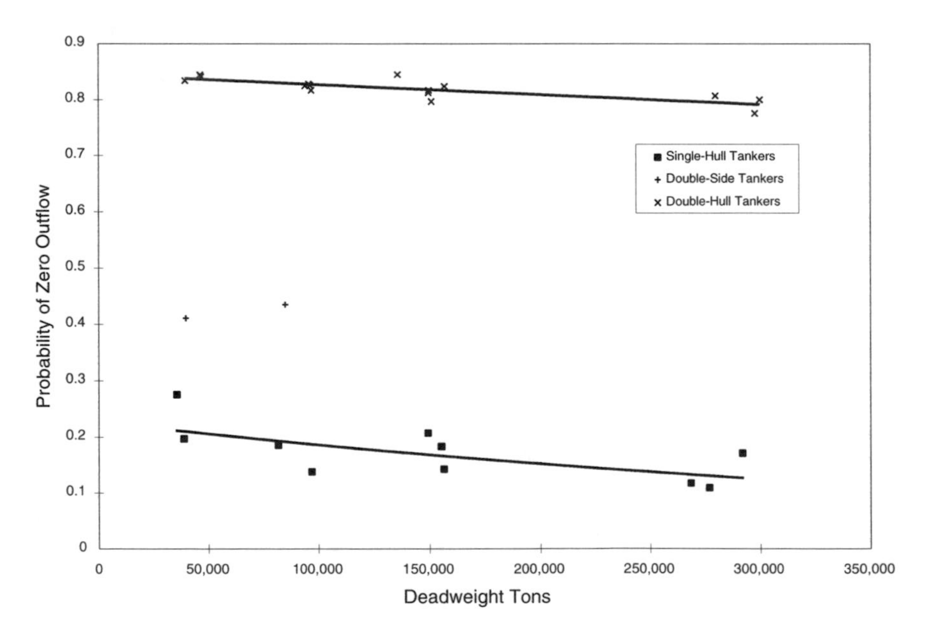

FIGURE K-14 Probability of zero outflow for single-hull and double-hull tankers.

increase outflow as compared to the more extensive cargo tank subdivision incorporated into the pre-MARPOL and MARPOL 78 designs. The light line on Figure K-15 represents a curve-fit of the single-hull mean outflow data. We find that the two double-side vessels evaluated in this study fall slightly above this trend line, indicating these double-side vessels will have comparable outflow volumes to the typical single-hull vessel. For double-side vessels with oil-tight longitudinal bulkheads, improved performance as compared to single hulls can be expected.

Mean outflow is influenced by the double-hull dimensions as well as the extent of internal subdivision within the cargo tanks. There is little variation in the arrangement of VLCCs, with most single-hull and double-hull designs incorporating a 5 × 3 cargo tank arrangement. Wing tank and double bottom dimensions for VLCCs typically fall between 3.0 and 3.5 meters. As a result, mean outflow values for VLCC are relatively consistent. In contrast, there is considerable scatter in the outflow values for tankers under 165,000 DWT. Figure K-16 shows the side and bottom damage contributions to mean outflow for the 150,000 DWT tankers evaluated in this study. The projected outflow is consistently lower for designs #150-D3, #150-D4, and #150-D5, all of which have an oil-tight centerline bulkhead over the length of the cargo block. Design #150-D1, with all single-tank-across cargo tanks, has the highest mean outflow. Design #150-D2 has an oil-tight centerline bulkhead arranged over about 40 percent of the cargo block, with single-tank-across cargo tanks arranged elsewhere. It is interesting

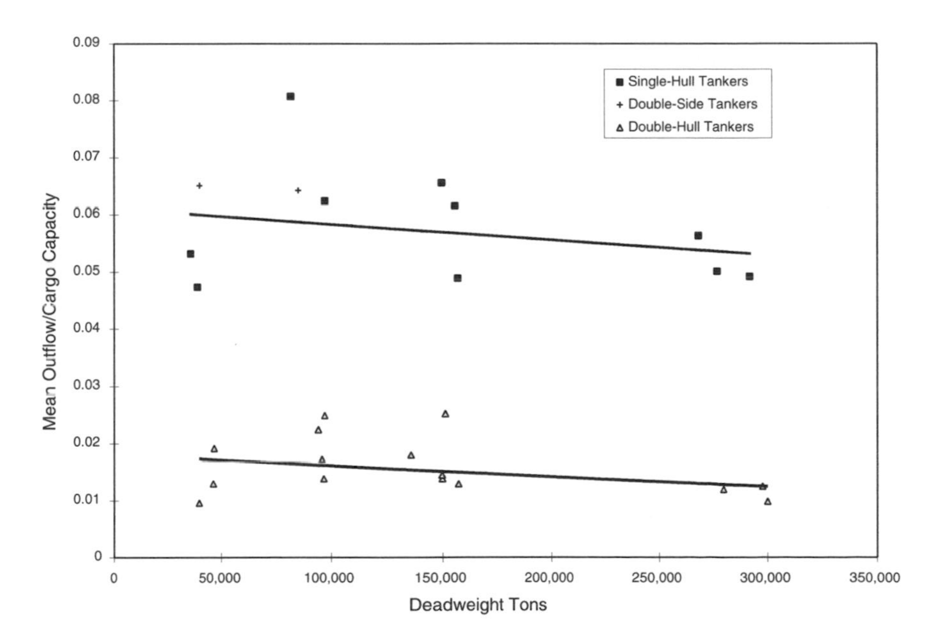

FIGURE K-15 Mean outflow for single-hull and double-hull tankers.

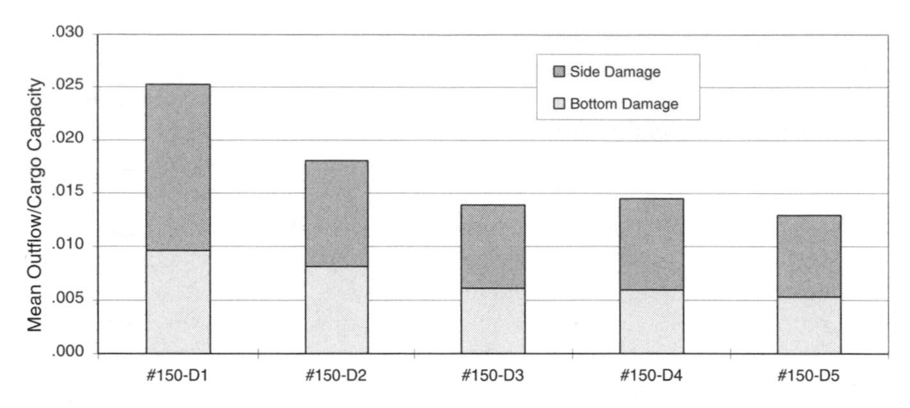

FIGURE K-16 Mean outflow data for 150,000 DWT double-hull tankers.

to note that the bottom damage outflow are relatively consistent, but the single-tank-across designs perform less effectively when subject to side damage. The closer spacing of transverse bulkheads on these designs increases the probability of breaching multiple cargo tanks. Once a cargo tank is breached, oil outflow is no longer limited to one side of the vessel.

As shown in Figure K-17, double-hull tankers without centerline bulkheads typically have twice the expected outflow of designs with oil-tight longitudinal bulkheads in way of the cargo block.

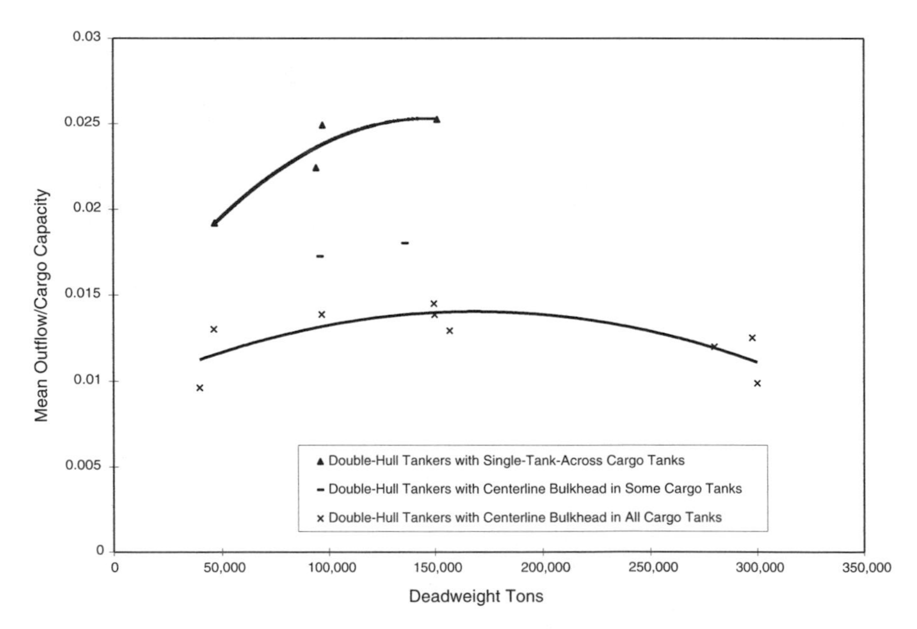

FIGURE K-17 Mean outflow for double-hull tankers with and without centerline bulkheads.

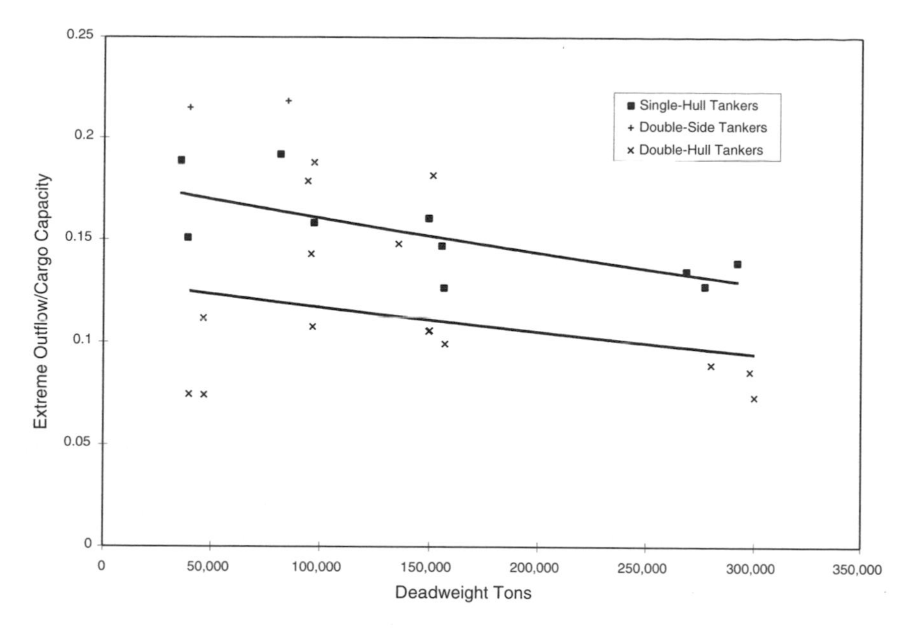

FIGURE K-18 Extreme outflow for single-hull and double-hull tankers.

Extreme outflow is a measure of a design's propensity to spill large volumes of oil in the event of a very severe collision or grounding. The extreme outflow parameters are plotted in Figure K-18. Whereas double hulls were shown to be 3 to 6 times more effective in avoiding spills and reducing mean outflow, double hulls are somewhat less effective in controlling large spills. There is considerable scatter in the data points, indicating that such parameters as internal subdivision and draft/depth ratio have a significant impact on extreme outflow. With regard to extreme outflow, the double-hull vessels with single-tank-across arrangements performed more poorly than both pre-MARPOL and MARPOL 78 vessels of comparable size.

The IMO Pollution Prevention Index E provides an overall picture of the outflow performance of a tanker. See Figure K-19 below. Single-hull tanker values generally fall between 0.3 and 0.4, whereas double-hull tanker values lie between .9 and 1.1. Sixty percent (9 of 15) of the double-hull designs had indices greater than 1.0, indicating equivalency to IMO's reference ships. In general, the ships with longitudinal oil tight subdivision in the cargo holds attained the highest indices. Of interest is design #150-D2, which has an Index E of 0.99, roughly equivalent to the IMO reference ship. Although approximately half the cargo oil capacity of this design is contained in single-tank-across cargo tanks, the detrimental effect of these tanks is offset by the contributions from the relatively wide wing tanks and deep double bottom tanks.

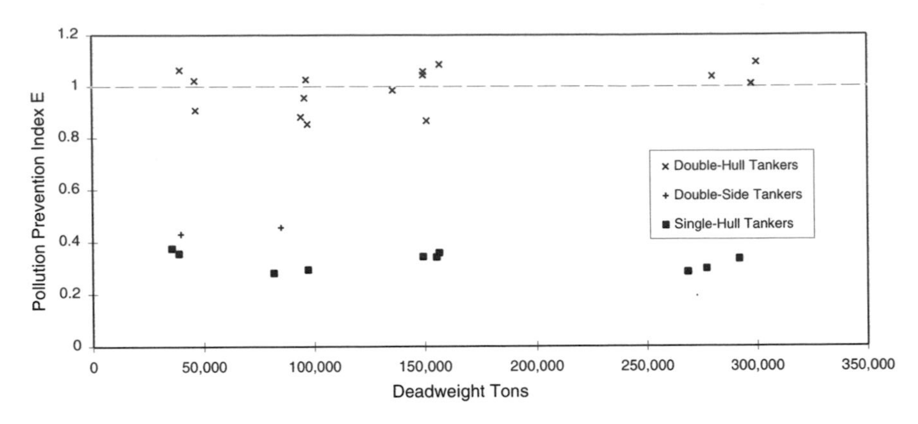

FIGURE K-19 IMO pollution prevention Index E for single-hull and double-hull tankers.

Observations on the Survivability of Tankers

There is no discernible difference between survivability characteristics of single-hull and double-hull tankers, with the survivability indices generally falling between 99 percent and 100 percent. Two of the ships in the 35,000 to 50,000 DWT range had values of 87.2 percent and 92.5 percent, respectively. However, these values are more heavily influenced by the level of compartmentation within the engine room and adjacent spaces than to the differences between single-hull and double-hull arrangements. For ships under 225 meters in length, MARPOL damage stability requirements do not require evaluation of conditions which breach the fore or aft engine room bulkheads. For certain designs, such damages result in nonsurvival conditions.

It should be noted that the the survivability index has been computed assuming a full cargo load, with all cargo tanks 98 percent full. Partial load conditions will likely have lower survival rates.

Observations on the Intact Stability of Tankers

With regard to intact stability, all single-hull designs are inherently stable. That is, for the worst possible combination of cargo and ballast tank loading, these vessels all maintained a GMt not less than 0.15 meters.

For the double-hull vessels, 73 percent (11 of 15) were inherently stable. The designs which have the potential of instability (#40-D1, #80-D1, #150-D1, and #150-D2) all have single-tank-across cargo tanks.

Designs #80-D1, #150-D1 and #150-D2 all had angles of loll below 8 degrees for the worst case loading situation, with no possibility of capsize. The load restrictions required to assure positive stability for these vessels are quite straightforward, requiring monitoring of any two ballast tanks. With all ballast tanks

2 percent full, the designs maintain positive stability through all possible cargo load conditions.

Design #40-D1 incorporates a single-tank-across arrangement for the cargo tanks and some U type ballast tanks. These tanks introduce large free surface effects when they are partially full. Also, the beam/depth ratio of 1.79 is relatively low. Although the vessel is in no danger of capsizing, an angle of loll of 16 degrees will occur for the worst case loading situation. This loll angle could be further increased if the vessel is asymetrically loaded due to efforts to correct heel through counter-balancing. The load restrictions to assure positive stability for this vessel are quite complex, requiring monitoring of both ballast and cargo tanks.

Observations on the Ballast Condition Analysis for Tankers

The double bottom and wing tank dimensions for existing double-hull tankers generally exceed the rule requirements, providing ballast capacity in excess of that required to achieve the minimum IMO drafts. All of the designs evaluated have forward drafts at least 19 percent deeper than the IMO minimum requirements, and most designs had drafts more than 50 percent in excess of the rule minimum.

Most of the double-hull designs evaluated have still-water bending moments in the ballast condition approaching the maximum permissible value assigned by the classification society. Exceptions are designs #40-D3 and #280-D3. Design #40-D3 has scantlings and consequently a permissible still-water bending moment significantly above rule requirements. Design #280-D3 has additional hull girder strength and deep ballast tanks located in the midships region.

As shown in Table K-25, the average double-hull design has a permissible still-water bending moment 9 percent in excess of the ABS standard value. This is 13 percent above the average for single-hull vessels analyzed. It should be recognized, however, that rule requirements for longitudinal strength have been liberalized since many of the single-hull tankers were built. Although the relative permissible bending moments are higher, it is possible that this may be a result of higher permissible stresses rather than increased structural strength.

TABLE K-25 Allowable Still-Water Bending Moments as a Percentage of the ABS Standard Value

	Single Hull	Double Hull
35,000–50,000 DWT Tankers	106	124
80,000–100,000 DWT Tankers	98	100
135,000–160,000 DWT Tankers	89	106
265,000–300,000 DWT Tankers	93	110
Average (all tankers)	96	109

Observations on the Oil Outflow Analysis and Survivability Analysis for Barges

Parallel to the findings for tankers, double-hulled barges exhibited significant improvements with regard to the likelihood of avoiding spills (larger values for the probability of zero outflow) and the mitigation of the amount of oil spillage (smaller mean outflow values).

Although the analysis for double-hull tankers did not extend to sizes below 25,000 DWT, it is expected that the mean outflow for tankers will be somewhat higher than for barges. This is because the reduced freeboard requirements for barges allow higher draft/depth ratios, which tends to reduce outflow from groundings.

It is important to remember that this study investigates the relative performance of a design to mitigate outflow, assuming that it has experienced a collision or grounding which breaches the outer hull. The overall outflow performance must also consider the likelihood that a given vessel will experience such an accident. Therefore, a comparison of barges and tankers cannot be made on the basis of the outflow parameters alone.

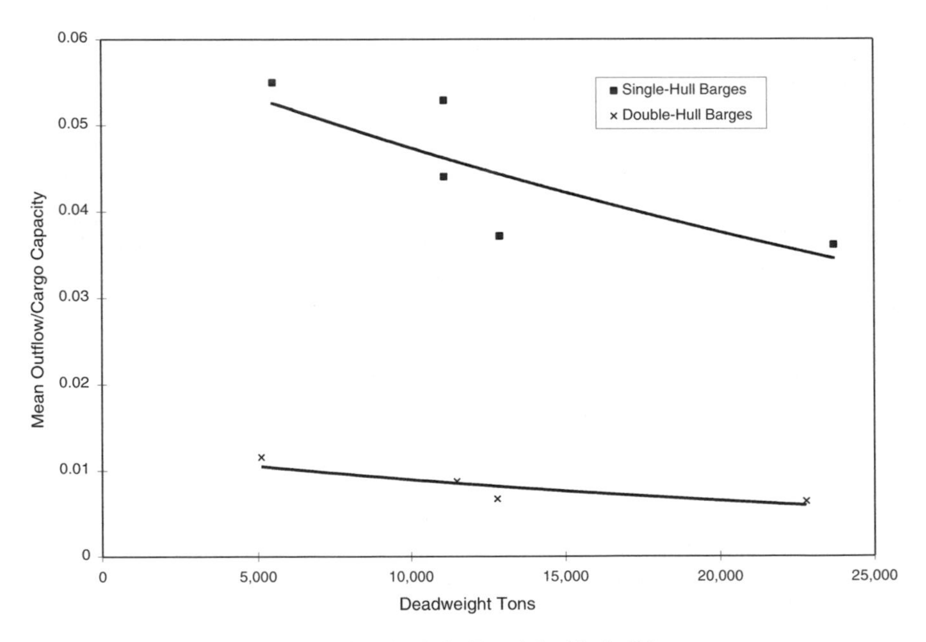

FIGURE K-20 Mean outflow for single-hull and double-hull barges.

Cautionary Notes on the Assumptions and Limitations of this Study

It is important to recognize that, due to both technical and practical limitations, there are many simplifications inherent in these calculations. The quantities of oil outflow do not represent a quantitatively accurate estimate of oil outflow, nor does the survivability index represent an exact determination of the probability that a certain design will survive a collision. Rather, these calculations provide a rational comparative measure of merit.

Some of the assumptions and simplifications in the development of damage case probabilities are:

- The IMO statistical database (Lloyd's, 1991) used for developing the probability density functions is based on 50 to 60 incidents involving tankers above 30,000 DWT.
- The probability density functions are "marginal" distributions. Locations, extent and penetrations are treated independently. Although some degree of correlation is expected, the correlated statistics are not currently available. It is believed that this approach is conservative in the sense that it tends to over-predict the amount of expected outflow.
- The historical casualty data primarily involve older, single-hull vessels. It is expected that extents of damage will be somewhat less for double-hull vessels.

The 19 double-hull vessels analyzed in this study represent about 5 percent of the double-hull tanker fleet operating today. Efforts were made to select representative vessels. However, there are some double-hull vessels built for specific trades which have quite different characteristics as compared to these representative vessels.

REFERENCES

American Bureau of Shipping. 1995. Rules for Building and Classing Steel Vessels. Part 3, Hull Construction, and Equipment. New York: American Bureau of Shipping.

David Taylor Research Center (DTRC). 1992. Summary of Oil Spill Model Tests. OTD 5/10, Annex 4. February. Washington, D.C.: U.S. Navy.

Herbert Engineering Corporation. 1996. HECSALV Salvage Engineering Software, Version 5.08. Houston, Tex.: Herbert Engineering Corporation.

International Maritime Organization (IMO). 1992. New Regulations 13F and 13G and Related Amendments to Annex I of MARPOL 73/78. London: IMO.

IMO. 1993. Explanatory Notes to the SOLAS Regulations on Subdivision and Damage Stability of Cargo Ships of 100 Metres in Length and Over. London: IMO.

IMO. 1995. Interim Guidelines for the Approval of Alternative Methods of Design and Construction of Oil Tankers under Regulation 13F(5) of Annex I of MARPOL 73/78. London: IMO.

Lloyd's Register of Shipping. 1991. Statistical Analysis of Classification Society Records for Oil Tanker Collisions and Groundings. Report No. 2078-3-0. London: Lloyd's Register of Shipping.

Michel, K., and C. Moore. 1995. Application of IMO's Probabilistic Oil Outflow Methodology. Paper presented at SNAME Cybernautics '95 Symposium. New York: SNAME.

National Research Council (NRC). 1991. Tanker Spills: Prevention by Design. Marine Board. Washington, D.C.: National Academy Press.

Society of Naval Architects and Marine Engineers (SNAME). 1988. Principles of Naval Architecture. Vol. I, Stability and Strength. Jersey City, N.J.: SNAME.

Tsukuba Institute. 1992. Model Tests by the Tsukuba Institute. Paper submitted to the IMO Marine Environment Protection Committee (MEPC). MEPC 32/7/1, Annex 6. London: IMO.

U.S. Department of Transportation (DOT). 1992. Alternatives to Double-Hull Tank Vessel Design, Oil Pollution Act of 1990. Report to Congress, Washington, D.C.: U.S. Department of Transportation.

APPENDIX L

Research on Double-Hull Vessel Technology since 1990

This appendix describes major research since 1990 aimed at enhancing understanding of the structural behavior of double-hull tank vessels, as well as improving ways to reduce potential outflow after an accident. Much of this research was planned or initiated before the promulgation of the Oil Pollution Act of 1990 (P.L. 101-380) (OPA 90), but the results have been obtained only during the last six years. The first section focuses on structural research on tankers. The ultimate objective of this research is the development of methods and data that will facilitate the design of hull structures with long life and good performance in accident situations. The second section addresses double-hull tanker design concepts that offer alternatives to conventional double-hull tanker construction.

STRUCTURAL RESEARCH

Major research projects in double-hull technology since 1990 have been conducted principally in the United States, Japan, the Netherlands, Denmark, and Norway. In 1991, Japan initiated a major seven-year structural research program on the prevention of oil spills from crude oil tankers under the Association for Structural Improvement of Shipbuilding Industry. Most of the research in the United States has been performed by the Massachusetts Institute of Technology (MIT), the Interagency Ship Structure Committee,[1] and the Society of Naval Ar-

[1]The Interagency Ship Structure Committee consists of the following member agencies: American Bureau of Shipping, Defense Research Establishment Atlantic (Canada), Maritime Administration, Military Sealift Command, Naval Sea Systems Command, Transport Canada, United States Coast Guard. The committee funds and pursues a research program to improve the hull structures of ships and other marine structures.

chitects and Marine Engineers. Structural research efforts in other countries have also been reported at international meetings, such as the International Ship and Offshore Structures Congress and the International Symposia on Practical Design of Ships and Mobile Units.

Structural Design Research to Reduce the Effects of Fatigue on Ship Life

As noted in Chapter 6, some problems with fatigue cracking of high tensile steels in double-hull tankers were encountered in the late 1980s. A double hull tends to be stiffer than its single-hull counterpart; this can affect residual stresses induced during construction and local stresses due to operational loads, both of which can result in initiation of fatigue cracks.

Advances in finite element stress analysis techniques have made it possible to obtain accurate and detailed stress estimates. For the most part, analyses of this type are now carried out routinely as an integral part of the design process by shipyards producing double-hull tankers and are no longer regarded as research studies. Structural details at welded junctions have been analyzed and redesigned to improve fatigue life. Experimental research in this area is under way to document the development of fatigue cracks in various joint designs. The large-scale tests conducted at the Krylov Shipbuilding Research Institute in Kiev, sponsored by Lloyd's Register of Shipping, represent one such example (Violette, 1995). The application of fracture mechanics to improve the fatigue life of ship hulls has already paid dividends. Nevertheless, there appears to be considerable potential for further progress in this area.

Structural Responses to Collisions

Since V.U. Minorsky's efforts in the late 1950s to correlate the interpenetration of colliding ships using accident data (Minorsky, 1959), there has been continuing research aimed at more accurately accounting for the structural details and approach characteristics of colliding ships. Although the early approach to predicting penetration largely depended on relatively simple energy accounting, the most recent methods are based on detailed analysis of plastic buckling, collapse, and fracture. The importance of postcollision ship motion and wave generation in the energy balance is now recognized. The evolving methods are applicable to all types of ship structures, including double hulls. The goals of this research are to allow the designer to evaluate the performance of competing designs in a variety of critical accident scenarios and to refine designs to meet specific performance goals.

Collision analysis has been greatly aided by modern nonlinear finite element methods, which have been increasingly used for research in this area during the past five years. Nonlinear finite element methods are now starting to be used to optimize double-hull designs with respect to plate thickness; steel strength; and positioning of inner and outer hull plates, side stringers, and transverse webs. Verification of analytical procedures using scale-model tests and actual collision data—where available—is a necessary part of the approach because of the inherent difficulty in modeling highly contorted collapse modes and the relatively crude criteria that are still employed to model plate- and weld-fracture during crushing.

Full-scale collision tests have been conducted using two inland waterway tankers, each of approximately 1,000-metric ton displacement, in a collaborative effort with support from a number of Dutch and Japanese groups (Vredeveldt and Wevers, 1992, 1995; Wevers et al., 1994). These tests were accompanied by detailed numerical simulations (Lenselink and Thung, 1992). A series of four impacts were conducted wherein one tanker fitted with a nominally rigid bow struck the other tanker's side at 90 degrees. Two of the impacts were against side sections of the ship having a single hull, whereas in the other two collisions the tanker was struck in side sections having a double hull. Data were recorded on penetration depth, collision force, strains in critical locations, and all six rigid-body motions of each of the ships. In addition, observations on cracking, which largely occurred along weld lines, were reported. The accompanying numerical simulations were successful in replicating major features of the collision, with the exception of crack patterns. The experimental data will be available for calibration of analysis methods in the future. Among the conclusions drawn from the joint Dutch-Japanese research were these: fracture initiation is dominated by the welds and is poorly characterized; the hydrodynamics of both ships during collision must be modeled correctly if penetration and collision forces are to be predicted accurately; and a sizable fraction of the energy dissipated in a collision goes into wave generation.

Structural Responses to Groundings

Most aspects of structural failure in tanker grounding incidents can be analyzed by the same methods used to analyze ship collisions. However, hull-girder failure (i.e., "breaking the back" of the tanker) and hull tearing are features specific to grounding that require specialized approaches. Hull-girder failures due to grounding have been examined with the aid of increasingly powerful numerical models within the last five years. Issues studied include whether dynamic effects contribute significantly to hull-girder collapse and the influence of friction between hull and seabed. The computational models are in reasonable accord with model and full-scale grounding tests, such as those undertaken in Denmark (Paik and Pedersen, 1995).

The U.S. Navy has conducted 1/4-scale model tests for strandings (loadings normal to the bottom of the hull) and groundings (combined normal and in-plane loadings) (Sikora and Bruchman, 1992; Melton et al., 1994; Rodd and Sikora, 1995; Sikora et al., 1995). These tests were part of a comprehensive program that also considered preliminary designs of double-hull vessels for both naval and commercial use, with particular attention to efficient producibility and long fatigue life. The grounding tests were accompanied by analytical work based on nonlinear finite element structural models similar to those used to analyze the Dutch-Japanese collision tests.

Initial efforts to assess the resistance of underside hull plates to tearing by a protrusion, such as a rock jutting up from the seabed, have been undertaken at MIT (Wierzbicki, 1995). This effort is couched within the framework of fracture mechanics, where the energy required per unit length in the tearing of a plate plays a central role in the analysis. The tearing energy for steel plate must be measured independently in a simulated tearing test. Then, the length of the underside rupture is estimated by accounting for the combined energy dissipated in the grounding from tearing and from plastic deformation of the hull during interaction with the protrusion, with due allowance for other mechanisms of energy loss. The mechanics of problems combining large amounts of plastic deformation and fracture are unusually challenging. Part of the difficulty in applications to ship hulls lies in the fact that the tearing energy constitutes a relatively small proportion of the total energy dissipated in the grounding, yet this energy is critical in determining of the extent of the tear.

The integration of a sound fracture analysis approach into collision and grounding analyses would constitute a major advance in the analytical tools available to assess and design double-hull tankers. The U.S. Navy has identified this as an important goal necessary to improve prediction capabilities for the design and residual strength analysis of ship structures against accidents and aggressive attacks. Observations suggest that most cracks will initiate and propagate in weldments. Criteria currently employed for the initiation of cracking during a collision or grounding are usually based on the attainment of some critical plastic strain locally in the weld or plate material. The validity of such criteria remain poorly established. Once a crack has initiated in a region of intense deformation, its subsequent spread requires a fracture mechanics analysis using the relevant fracture properties of the weldment or plate. The joint Dutch-Japanese study cited above found that cracks that formed and propagated outside the immediate penetration region had to be accounted for if accurate predictions for collisions or groundings were to be achieved. Observations concerning the tendency for cracks to initiate in welds highlight the importance of weld quality.

Structural design approaches used today ensure that tankers have sufficient strength to withstand the loads encountered in regular operation, but there are no provisions for the loads encountered in accidents. Similarly, the outflow performance of tankers is based on tank subdivision only, and no consideration is given to the performance of the structure in collisions and groundings. The development of tools that could be used to design tanker structures for good performance in accident situations will provide an important advance in the design of tankers. The research described above has this objective, although much work is still required before the results of research efforts can be translated into practical design tools.

Other than work being conducted by the U.S. Navy, the research of Wierzbicki and his coworkers at MIT represents the main activity in the United States dedicated to the development of advanced analysis methods for ship structures along the lines indicated above. Nearly all of Wierzbicki's effort is supported by industry sources, mainly from abroad. The committee is concerned that important research opportunities may be missed due to the absence of any significant U.S. agency funding for work on the development of analysis tools for structural integrity of ship structures. Most of the powerful computer codes used to analyze the nonlinear deformation of structures have been developed in this country, and the expertise needed to extend them to include effects such as collapse and fracture also resides in this country.

ALTERNATIVE DOUBLE-HULL TANKER DESIGNS

Several design concepts have been developed since 1990 that offer alternatives to "conventional" double-hull tanker construction. The concepts can be divided into three major categories: (1) designs to improve producibility, (2) designs to improve outflow performance, and (3) designs to reduce maintenance costs.

Improved Producibility

A number of designers have proposed unidirectionally stiffened double-hull tanker designs in which the amount of transverse structure has been minimized to maximize productivity in construction. This concept was proposed before double-hull tankers became mandatory (Okamoto et al., 1985). The unidirectionally stiffened structure improves construction productivity by reducing the number of structural joints and by allowing maximum use of automatic welding. A disadvantage of this design is that a unidirectional structure requires smaller tanks than a conventional tanker structure, thus increasing the subdivisions and the weight of steel in the vessel. The unidirectionally stiffened double-hull concept has been applied to small product tankers built in the Far East but has not been applied successfully to larger tankers to date.

The U.S. Navy has undertaken a study of a unidirectionally stiffened double-hull design (advanced double-hull concept) with emphasis on fatigue life and producibility as well as resilience to collision, grounding, and attack (Melton et al., 1994; Sikora et al., 1995). The MarC Guardian tanker project—supported by the Carderock Division, Naval Surface Warfare Center, and Advanced Research Projects Agency (ARPA) Maritech—is proposing a unidirectionally-stiffened design concept that uses slightly curved plating for the outer and inner hulls of a vessel, thereby eliminating the need for local plate stiffeners. The spacing between hulls, the spacing of longitudinal girders, and the spacing of transverse structures can be standardized for various vessel sizes by using curved plates (Goldbach, 1994).

Improved Outflow Performance

Five European shipbuilders—Astilleros Españoles, Bremer Vulkan, Chantiers de l'Atlantique, Fincantieri, and Howaldtswerke Deutsche Werft—have cooperated in designing an ecological tanker concept called the "E3 tanker" (Paetow, 1992). The name refers to several tanker designs that provide varying levels of protection against oil spills. The arrangement and construction of a "standard E3" concept do not differ from a typical very large crude carrier (VLCC) double-hull tanker design. However, the "superecological E3" tanker design has small cargo tanks and double-hull dimensions that were optimized to reduce the probability of oil spills using statistical data on damage extent and damage locations on the hull. So far, one standard E3 tanker has been built in Spain (Gutierrez-Fraile et al., 1994).

Reduced Maintenance

NKK Corporation and World-Wide Shipping Agency have proposed an alternative double-hull design concept in which the double-hull spaces are dry void spaces and ballast tanks are arranged in the inner hull in a manner similar to that in a single-hull tanker. The concept aims to eliminate concerns associated with the operation and maintenance of double-hull tankers. The increased initial cost is offset by lower maintenance costs (Akita et al., 1995). Such a design would have to be larger overall to provide for adequate ballast in addition to void tanks.

All of the double-hull tanker designs described above are still at the concept stage, with the exception of small unidirectionally stiffened tankers built in the Far East and the standard E3 tanker built in Spain. They have not yet been proven to be competitive alternatives to conventional double-hull tanker designs.

REFERENCES

Akita, T., K. Kitano, Y. Sumikama, H. Tsukuda, M. Toyofuku, K. Shibasaki, J. Hah, and K. Furukawa. 1995. A revolutionary design of double-hull oil tanker. Proceedings, Offshore and Polar Engineering Conference (ISOPE-95), The Hague, June 11–16. Golden, Colo.: International Society of Offshore and Polar Engineers.

Goldbach, R.D. 1994. MarC Guardian tanker concept—Introduction of a world competitive American environmental tanker. SNAME Transactions 102:265–294.

Gutierrez-Fraile, R., H. Rosemberg, P. Terson, A. Cumin, and K. Paetow. 1994. The European E3 tanker—Development of an ecological ship. SNAME Transactions 102:237–264.

Lenselink, H., and K.G. Thung. 1992. Numerical simulation of the Dutch-Japanese full-scale ship collision test. Pp. 771–785 in Proceedings of the First Conference on Marine Safety and Environment, Ship Production, Delft, June 1–5. Delft: Delft University Press.

Melton, W., J. Beach, J. Gagorid, D. Roseman, and J. Sikora. 1994. Advanced double-hull research and development for naval and commercial ship application. SNAME Transactions 102:295–323.

Minorsky, V.U. 1959. An analysis of ship collisions with reference to protection of nuclear power plants. Journal of Ship Research 3(1):1–4.

Okamoto, T., T. Hori, M. Tateishi, S. Rashed, and S. Miwa. 1985. Strength evaluation of novel unidirectional-girder-system product oil carrier by reliability analysis. SNAME Transactions 93:55–77.

Paetow, K.H. 1992. The E3 tanker: Design-structure-ballast tank protection. Presented at the Tanker Structure Cooperative Forum Shipbuilders Meeting, London, October.

Paik, J.K., and P.T. Pedersen. 1995. Collapse of a ship's hull due to grounding. Pp. 75–80 in Volume I, Proceedings, International Conference on Technologies for Marine Environment Preservation (MARIENV 95), Tokyo, September 24–29. Tokyo: The Society of Naval Architects of Japan.

Rodd, J., and J. Sikora. 1995. Double hull grounding experiments. Proceedings, Offshore and Polar Engineering Conference (ISOPE-95), The Hague, June 11–16. Golden, Colo.: International Society of Offshore and Polar Engineers.

Sikora, J., and D. Bruchman. 1992. Pp. 89–99 in Proceedings, Offshore and Polar Engineering Conference (ISOPE-92), San Francisco, June 14–19. Golden, Colo.: International Society of Offshore and Polar Engineers.

Sikora, J., R. Michaelson, D. Roseman, R. Juers, and W. Melton.1995. Double-hull tanker research—Further Studies. SNAME Transactions, preprint paper no. 15. Jersey City, N.J.: Society of Naval Architects and Marine Engineers.

Violette, F.L.M. 1995. Lloyd's Register integrated fatigue design assessment system. Pp. 242–249 in Proceedings, International Conference on Technologies for Marine Environment Preservation, Volume 1, Tokyo, September 24–29. Tokyo: The Society of Naval Architects of Japan.

Vredeveldt, A.W., and L.J. Wevers. 1992. Full-scale ship collision tests. Pp. 743–769 in Proceedings, First Conference on Marine Safety and Environment Ship Production, Delft, June 1–5. Delft: Delft University Press.

Vredeveldt, A.W., and L.J. Wevers. 1995. Full-scale grounding experiments. Pp. 11–112 in Proceedings of Conference on Prediction Methodology of Tanker Structural Failure and Consequential Oil Spill, Tokyo, April. Tokyo: Association for Structural Improvements of the Shipbuilding Industry in Japan.

Wevers, L.J., J. van Vugt, and A.W. Vredeveldt. 1994. Full-scale six degrees of freedom motion measurements of two colliding 80 m long inland waterway tankers. Pp. 923–930 in Proceedings of the 10th International Conference on Experimental Mechanics, Lisbon, June 18–22. Rotterdam: A.A. Baldema.

Wierzbicki, T. 1995. Concertina tearing of metal plates. International Journal of Solids and Structures 32(19):2923–2943.

APPENDIX

M

Summary of Questionnaire Responses from Owners and Operators of Double-Hull Tank Vessels[1]

I. Operation of double-hull tankers

1a. What is your experience with operational safety of double-hull (DH) tankers in regard to:

Stability during loading and discharging

A. No stability problems. It is important to build DH vessels with center bulkheads.
B. No stability problems.
C. Not perceived as a problem. Officers must be aware of the limitations.
D. Modifications to generic tanker specifications necessary for the company's special cargo trade. Structure added into center tanks. "Caution posters" displayed in Cargo control room; information on any restrictions documented when duties are handed over.
E. No stability problems. However the Trim and Stability Booklets including any restrictions must be complied with. Stability of DH hull tankers is an issue. Early designs largely ignored the trade-off in intact and damage stability characteristics.
F. No significant problems. DH tankers have centerline (CL)heads that reduce free surface moment. OBOs (Oil-bulk-ore [vessel]) do not have CL bulkheads: Masters cautioned of hazardous stability conditions.
G. Special precautions have to be taken with regard to stability.

[1]The responses are labeled A, B, C, ... in accordance with individuals who provided them. When a letter is missing, that individual did not respond to the question.

H. Satisfactory experience. Design requirement for vessels to remain stable during loading or discharge. Concerned about stability problems for ships without longitudinal bulkheads.
I. No specific problems. Officers and crew well trained in cargo operation of double hulls. Sizing ballast pumps in relation to cargo discharge capability important.
J. Problem maintaining stern trim due to discharging patterns in parcel chemical tanker trade.
K. This is an issue but can be dealt with easily.
L. Free surface effect has increased dramatically.
M. No problems with company tankers. A problem for tankers built without a centerline bulkhead in cargo tanks and DB (double bottom) tanks.
N. Stability problems during ballasting and de-ballasting on a 90,000-deadweight ton (DWT) tanker with no centerline bulkhead in DB tank; operational procedures required.

1b. What is your experience with operational safety of double-hull tankers in regard to:

Safe access to ballast spaces

B. Access through hatches and inclined ladders strictly controlled. Procedures follow Chapter 10 of the International Safety Guide for Oil Tankers and Terminals (ISGOTT). Horizontal stringers and larger longitudinals provide access in tanks.
C. To guarantee safe access, warning signs have to be posted.
D. Access through openings on main deck using steel ladders. Number of bays without access opening minimized. Horizontal decks provide access for inspection and maintenance. Large openings in intermediate decks for direct access to other levels.
E. Stringent safety regulations in effect: no problems experienced. Classification requirements for design do not provide good access for inspection. In the absence of permanent access some inspection methods (rafting, etc.) are more difficult in DH spaces.
F. Complexity of structure may increase safety risks compared to single-hull (SH) tankers. More openings on deck may be required for adequate ventilation.
G. Access to ballast spaces good.
H. Satisfactory experience. DH vessels designed with built-in walkways and adequate access. Space entry procedures enforced.
I. Vertical ladders in lieu of inclined ladders: not unsafe but harder to use. Independent rescue hatches in every tank for direct access to main deck in case of emergency.

J. People should not go into ballast spaces unless necessary. Current requirements for access sufficient.
K. Same care required as on entry into any ballast space.
L. More difficult access; longer distance to escape; complex construction requires knowledge of the configuration.
M. Only one entry into ballast tanks: access both fore and aft convenient but not necessary. Large openings for emergency access require provisions to prevent falls.

1c. What is your experience with operational safety of double-hull tankers in regard to:

Ventilation of ballast spaces

A. Portable fans at hatches. Flexible hoses if needed.
B. Ventilation through ballast lines. Air supply from inert gas main connected to a flexible hose. Ventilation time (4,000 m^3 in DB and 5,000 m^3 in side) on order of 3.5 to 4.5 hours. Alternative: fill tanks with water and then empty them.
C. Ventilation via airpipes at forward and aft end of tanks, if necessary using portable water-driven fans.
D. Ventilation by mushroom ventilators. Opening sizes adequate for air intake and exhaust.
E. Flooding relief vent head and hatch opening for natural vent during ballast or de-ballast operation. Purge pipes from deck to centerline bulkhead. Mechanical ventilation and vapor testing prior to entry. Forced ventilation to double bottom difficult.
F. See response to 1b.
G. Ventilation of ballast spaces good.
H. Satisfactory experience. Sufficient venting facilities provided. Ventilation on DH ships requires more attention than on SH vessels.
I. Difficult to ventilate ballast tanks. Cross-connection from inert gas line to ballast line can provide good circulation even in DB tanks.
J. Can be done adequately with existing fixed and portable units. Testing and "safe entry" procedures are important.
K. As in any other vessel design.
L. Pockets without oxygen may exist. In case of oil leakage, pockets of flammable gases may exist even after ventilation.
M. Flexible hose used to provide air; discharge through the tank opening. Safety always a major concern. Risk of cargo leakage to DH spaces overemphasized: must always be vigilant. Instrumentation may lead to complacency.

1d. Any other safety issues that need to be addressed.

A. DH tankers are always capable of ballasting to a safe draft. Inerting of ballast tanks possible using flexible hose connections to inert gas plant.
B. Nonc
C. It is important to have a capability to inert all ballast and void spaces using emergency connections to the ballast pipe in the double bottom.

D. Modifications made: ballast pump designed to trip when tank 98 percent full; hydrocarbon detection sensors in ballast spaces; inert gas system (IGS) can be used for forced ventilation; portable probes to test ballast tank atmosphere and treat water for microorganism.
H. Must have means to promptly detect hydrocarbons in ballast spaces and identify structural problems.
I. Accumulation of sediment in double bottom requires good drainage (design requirement). Fixed IGS (for ballast spaces) should be a requirement.
J. Product tankers should not be allowed to carry noxious liquid substances (personnel do not have proper experience, and construction standards are insufficient).
L. Permanent IG piping should be considered for all segregated ballast tanks (SBTs) and void spaces in case of oil leakage. More stringent check of tank atmosphere prior to entry required.
N. No difference in shiphandling.

2. Are there significant differences in cargo operations between double-hull and single-hull tankers?

A. DH tankers are more flexible in ballasting and de-ballasting: More operational flexibility. Can discharge part cargoes in almost any sequence.
B. Cargo operations generally easier on DB tankers: stripping and cleaning easier. Stripping can normally be done by main cargo pumps and time for tank washing is reduced.
C. Cargo tanks on DH tankers easy to discharge and clean. Unloading operation is faster on DH tanker or OBO (quicker discharge, tank washing, and stripping).
D. Enforced on DH tankers: understanding of arrangement (subdivision, piping, etc.) ; monitoring and constant awareness of stability and restrictions in concurrent ballast-cargo operations; listing moments; shear and bending stresses; cargo heating control.
F. No significant differences. Cargo operations easier on DH tankers.
G. Stability has to be carefully monitored during cargo operations.
H. DH tankers have better cargo outturn and tank washing characteristics.

I. No
J. No difference in cargo operations. Significant differences in design, maintenance and life expectancy (depends on planning and maintenance).
K. Monitoring of stability.
L. Accurate stability calculations required prior to and during cargo operations. No ballast in cargo spaces.
N. Discharge of DH ships better than SH ships.

3. Have you established operational procedures specifically for double-hull tankers?

A. Cargo and ballast operations on double-hull tankers are comprehensively described in vessels' operational procedures.
B. Monitoring of ballast tanks emphasized. During loaded passage, all ballast spaces are monitored weekly using portable gas detectors and checked with a sounding rod. Visual check after ballasting. All checks recorded.
C. No
D. DH tankers: ventilation of ballast spaces; stability instructions; IGS operation for ballast spaces; restrictions in ballast or cargo handling documented for "hand-over"; coating inspection and maintenance; detection of hydrocarbons in ballast spaces.
E. No special operating procedures established except requirements in Trim and Stability Booklets for ballasting and cargo handling sequence.
F. No; except cautionary advice as required.
G. Procedures for cargo operations with regard to stability.
H. Existing operational procedures adopted from those for DB tankers.
I. No
J. No. All ships have double bottoms and/or double sides.
K. Yes, to address stability with free surface in cargo tanks.
L. Stability procedures and procedures in case of oil leakage to ballast tanks have been established.

II. Inspection and maintenance of double-hull tankers

1. Please provide information on structural and tank coating inspection frequencies and practices on double-hull tankers.

A. Ballast and cargo tanks are inspected at least once a year.
B. Each laden voyage, ballast tanks inspected. Coating and structure inspected in two tanks each laden voyage (i.e., all tanks inspected every year). Minor coating repair during inspection.

C. Crew inspects ballast and void spaces every three months and repairs paint damage when necessary. Detailed inspection by an independent surveyor approximately every 2.5 years.

D. Coating and structural inspection at least once a year by technical inspector. Safety inspection every 120 days. "Guidelines for Enhanced Survey" and "Standard Coating Condition Inspection Guidelines" followed. American Society for Testing and Materials (ASTM) rust grade principles applied.

E. Crew inspects ballast tanks annually. Outside contractors monitor coating and structure on a schedule that follows survey schedules: new vessels, five-year cycle; older vessels, 2–3 year cycle.

F. Structural and coating inspection of coated tanks every other voyage. Ballast tanks inspected at least every other month.

G. Ballast and cargo tanks inspected annually by superintendent or ships' officers and by classification society as required.

H. All tanks are inspected on six-month schedule.

I. Visual inspection every three months.

J. Structure and coatings inspected every six months.

K. Nearly 100 percent sound coating should be maintained.

L. Coating inspection every six months. Possible damage repaired after each inspection.

M. Eggcrate-type structure improves quality of inspection on DH ships. Areas not easily accessible inspected using video camera, portable staging, and rafting. If additives are used for mud removal, surfaces become very slippery.

2. What is your experience with different types of coating in ballast spaces? Have you encountered significant corrosion problems? If so, please describe.

A. 27,000 DWT DH tankers since 1988: routine maintenance during yearly inspections; no significant corrosion problems. Coating in ballast tanks light colored. One of the 299,000 DWT DH tankers has had coating damage due to poor work at the yard.

B. Tried soft coatings with limited success. Proper protection with epoxy paint system (properly formulated and applied). No severe corrosion problems in coated, regularly inspected or maintained tanks.

D. High built coal tar epoxy coating. Good surface preparation essential. Experience limited to four years; no significant breakdown or corrosion. Quality control during construction is key to good coating.

E. Five vessels built in 1970s required considerable attention between 15 and 20 years of age. Vessels that carried heated cargo had far worse failure rates than others. Some coating failure on new vessels due to poor quality control during construction.

F. Coal tar epoxy used. No major corrosion problems; touch-ups made as required.
G. Tar epoxy. No special problems encountered.
H. Corrosion in ballast spaces of SH and DH vessels wherever coatings not properly maintained.
I. Used both coal tar epoxy and light-colored epoxy systems. Sporadic failure of coal tar systems on two vessels.
J. Experience with coatings from soft tar to pure epoxy. Soft epoxy tar-type, poor for wet spaces; pure epoxy, best. Significant corrosion in DH spaces in the past. Corrosion problems contributed to scrapping six ships in past five years.
K. Careful initial coating application and adequate coating thickness give effective corrosion control.
L. Prefer to answer question after company has more experience with OBOs built 1992–1994. Coating to date in excellent condition. No corrosion.
M. Importance of surface preparation emphasized. Continuous coating inspection and maintenance is key to success. Light-colored coatings preferred. Spot maintenance extended life of coating to 15 years.
N. Combination of epoxy coating and anodes gives a long life.

3. What are your current practices with regard to ballast tank coatings (include type, number of coats, thicknesses)? From your experience, what is the expected life of the coatings?

A. 27,000 DWT, built 1986—2 × 150 microns of coal tar epoxy; 299,000 DWT, built 93/95—1 × 150 microns surface tolerant epoxy, 1 × 150 microns modified tar epoxy (light color); with proper maintenance and initial application, coating will last vessel's life.
B. Surface preparation and two coats of tar epoxy. Stringent inspection and quality control during building. Effective lifetime of coating system 15–20 years.
C. Coal tar epoxy 2 × 125 microns plus anodes (pitguard anodes on bottom). With proper maintenance painting system lasts vessel's life. Next generation of OBOs will use light-colored coatings for ease of inspection.
D. High built type coal tar epoxy. DFT minimum 200 microns. Aluminum and/or zinc anodes. With ideal surface preparation, 25 years; realistically, 15 years.
E. Experimenting with coating suppliers and blasting methods. Primary method for large-scale maintenance work: dry grit blast, dehydrating, and coating with Devoe 235 Epoxy, two coats plus stripe.
F. Coating thickness of coal tar epoxy 250–300 microns. Expected life approximately 10 years.

G. 2 × tar epoxy, each 125 microns. Grinding of edges and strip coating. Life expectancy 10–20 years depending on workmanship.
H. Two coats and two stripe coats of light-reflecting, light-colored, modified epoxy (not coal tar base), minimum 250 microns total DFT. 100 percent anode system. With normal maintenance, life expectancy at least 15 years, provided quality work at construction.
I. Light color epoxy 2 × 150 micron dry film thickness. Expected life 10 years.
J. Minimum three coats (100–150 micron thickness each) pure epoxy. Life expectancy 20 years with minimal maintenance.
K. Tar epoxy, 500 microns in three coats. Maintenance and cathodic protection extend life of coating to useful life of ship.
L. Surface preparation to SA 2.5, coal tar epoxy, 2 × 125 microns, plus three stripe coats. Expected life 15–20 years.

4. Do any of your maintenance and inspection practices for single-hull tankers differ from those used on double-hull tankers?

A. No
B. More effort in monitoring ballast tanks on the first-generation DB tankers.
C. No
D. Inspection on DH tankers simple compared to SH tankers. More stress concentrations in DH tanker structures due to higher rigidity; more frequent inspection warranted.
E. Coating inspection and maintenance more critical in DH vessels: high cost of coating replacement, low life-time expectancy for replaced coating, if steel replacement required, may force vessel to early retirement.
F. No
G. Operates only double-hull tankers.
H. Periodic inspection and maintenance of DH spaces.
I. Flushing of sediment from double bottoms. Increased inspection requirements.
J. Not applicable.
K. Maintenance and ballast tank coatings.
L. Same as far as coated tanks are concerned.
N. Cost of maintenance less for DH than SH tankers due to improved accessibility (egg-crate structure).

III. Design of double-hull tankers

1. Have you had any structural problems on double-hull tankers?

A. Heavy weather damage on a very large crude carrier (VLCC). Ballasting more forward prevented similar problems.
B. So far no structural problems.

C. A small number of leakages into upper stool spaces due to faulty welds.
D. No significant problems (five-year operational experience).
E. No structural problems on new tankers. On older vessels, coating more of a problem than structure.
F. No major problems. Detail design and welding sequence important.
G. No problems.
H. No problems to date.
I. No problems in newbuild DH tankers. Company strongly supports efforts of American Bureau of Shipping (ABS) to modernize ship structure analysis.
J. No significant problems.
K. None.
L. So far no problems, but complex design details increase possibility of fractures.
M. Some structural details are areas subject to fracture. Structural modifications have been carried out.
N. Two minor incidents of leakage into double hull; three minor incidents of leakage into double bottom.

2. What is your experience with high-strength steel construction?

A. No particular problems.
B. Good experience. Attention paid to details and workmanship. Too often necessary to increase scantlings above rule requirements. Buckling and fatigue criteria checked. Careful in defining loading conditions.
C. No problem as long as good coating system protects against corrosion.
D. Excellent experience. Corrosion protection key to long life: reduction of plating thickness from 19–32 mm in 1970s to 17–19 mm enforces this. Exposure to excess heat should be prevented. Dedicated crew and committed technical support important.
E. No experience worthy of comment. New vessels built with approximately 70 percent high tensile steel (HTS), limited to Grade 32.
F. Limited experience. Company believes in use of mild steel. High tensile steel kept to a minimum in newbuildings.
G. Cracks in high tensile side shell longitudinals between bilge keel and ballast load line.
H. Localized high stresses and fatigue can lead to accelerated corrosion and cracks. Location, amount, and type of HTS and shipbuilder's experience with HTS important.
I. Company's ships have larger percentage of mild steel than other tankers. DH fleet too young to show problems associated with high-strength steel.
J. Currently not using high-strength "black" steel in new designs.
K. May not be available everywhere for repairs.

L. High-strength steel construction requires good surface preparation and good-quality coating, as scantlings are reduced.
M. Experienced high number of cracks in 165,000 DWT class due to use of high tensile steel.

3. What design changes would you suggest in future double-hull tankers?

A. None
B. Normally minimum class requirements for longitudinal bending moments insufficient. Deflections of secondary members important. Ballast tank amidships to reduce bending moment is a possibility. Details according to Tanker Structure Cooperative Forum (TSCF) recommendations.
C. Better accessibility to ballast tanks and light color paint in ballast spaces.
D. High-tensile steel should be restricted to internal structure. Reduction in steel thickness if coating thickness increased not to be allowed. Certification of outer shell structural welding inadequate and inconsistent. Surface preparation rules needed.
E. Standards for access, staging fittings, coatings should be mandatory rather than at option of owner.
F. Double hulls may be useful for smaller tankers (order of 45,000 DWT), less so for larger vessels. Perhaps a gradual double hull replaced by double bottom for larger vessels.
H. Requirement for inherent positive stability throughout ballast or cargo handling; light-colored coatings; high-volume, continuously monitoring hydrocarbon detection system; shipyard design and practices to be certified by class if high tensile steel used
I. Requirements for redundancy, alarm, and automatic changeover for steering gear in event of single failure. Increased powering requirement. Requirement for emergency propulsion.
J. Design specific to trade and size. What is applicable for an ultralarge crude carrier (ULCC) or VLCC is not applicable to smaller ships. In general, U and L tanks should be avoided for small size ships.
K. None
L. Longitudinal center bulkheads would improve stability.
M. Easy access for inspection should be included in structural design. Coating regulations, which inhibit development, should not be established.

IV. Fleet Information

1. Based on your experience, what are the advantages of double-hull tankers compared to single-hull tankers?

A. Always capable of ballasting to safe draft for immediate departure in case of emergency.

Capable of ballasting to heavy weather ballast without cargo tanks.
No water in cargo tanks—practically no corrosion; increases expected lifetime.
Easy access to frame structures that are mainly in ballast spaces.
Almost complete discharge of cargo.
Easy tank cleaning.
Increased environmental protection.

B. Added protection against cargo outflow in case of low-impact casualty.
Efficient stripping and tank washing; good cargo turnout.

C. Faster cargo unloading (discharging, tank washing, stripping)

D. Psychological shield in low-impact groundings. (However, due to structural rigidity may cause fracture of shell plating.)
Politically acceptable design.

E. Greater cargo outturn.
Fewer tank washing machines.
Greater protection from minor contact damage or oil spill.

F. Easier to load and discharge.
Good protection in low-impact collisions and groundings.

G. Safety in groundings or collisions.
Easy to clean cargo tanks.
Easy to empty cargo tanks.

H. Pollution protection for certain types of casualties.
Better cargo outturn and pumping performance.
Superior tank washing results.
Better access to inspect ballast tank structure.
Meets legal requirements.

I. Safer than single hulls.
Cleaner than single hulls.

J. Regulations will hopefully force scrapping of older ships.

L. Greater SBT capacity.
Reduced risk of pollution in case of grounding or collision.
Reduced risk of pollution.
Better heating performance.
Better stripping ability.

M. Eggcrate structure in way of side shell and bottom structure resistant to fatigue- related failures.
Inspection-friendly structure if intermediate stringers provided in wing ballast tanks.

N. Double hulls eliminate piping leaks as major source of pollution (*no cargo pipes in ballast tanks*).

2. Based on your experience, what are the disadvantages of double-hull tankers compared to single-hull tankers?

   A. Larger, lightweight, beam and draft.
   More expensive to build.
   More expensive canal and port expenses.
   Today's market offers no compensation for higher costs of DH tanker.

   B. Ballast tanks have large surfaces coated with sophisticated and expensive coating: need continuous monitoring and maintenance.
   Cleaning of ballast space after a possible leakage.
   Higher building cost.

   D. Excessive cost for no gain in safety or environmental preservation.
   Reduced cargo capacity.
   Increased ballast (non-earning).
   Increased port dues and insurance costs due to increased gross registered tonnage (GRT).
   Increased coating areas in ballast spaces.
   Heavier (not necessarily stronger) hull structure.
   Potential for hydrocarbon leakage to ballast spaces. Potential for explosion.
   Increased longitudinal forces.
   Increased transverse free surface.
   Poor accessibility for inspection and maintenance in double bottom.
   Poor initial, static, and dynamic stability.
   Extra maintenance costs.
   Structure will not withstand forces due to collision or grounding.
   Alternate design should be considered.
   No return on higher cost. Most oil majors continue to embrace substandard tonnage at low freight cost.

   E. DH vessels need more resources to properly manage them. Inadequate coating maintenance, structural problems if vessels built to class rules only, and stability problems may lead to problems for the industry.
   More critical stability.
   Higher construction cost.
   Ballast tank coating critical issue.
   Ballast tank ventilation difficult.
   Difficult to salvage after hard grounding.
   Greater beam or freeboard.

   F. More equipment required to monitor void spaces.
   Reduced cargo carrying capacity.
   More surfaces to maintain.
   For larger vessels, not much advantage in way of environmental protection.

G. Stability problems if no centerline bulkhead.
   Problems if leakage in inner hull.
H. Explosion risk in double-hull spaces if vapor detection system not fitted.
   Increased construction and maintenance cost.
   Stiffer hull structure may lead to localized cracking.
   Increased vigilance required to ensure integrity of double-hull spaces.
   Increased port and insurance costs due to greater GRT.
I. Approximately 10 percent increase in cost.
L. Stability problems with center bulkhead.
   Risk of fractures and oil leakage into SBT could create dangerous atmosphere.
   Difficult to clean, gas free, and repair.
   Possible risk of local steel wastage or loss of strength due to deteriorated coating.
N. Cleaning mud from ballast spaces a bigger problem than on SH ships.

Acronyms and Glossary

ACRONYMS

13F	Regulation 13F of Annex I of MARPOL 73/78
13G	Regulation 13G of Annex I of MARPOL 73/78
ABS	American Bureau of Shipping
ADR	Alaska Department of Revenue
AIMS	American Institute of Merchant Shipping
ANS	Alaskan North Slope
APCIS	Asia-Pacific Computerized Information System
ARPA	Advanced Research Projects Agency
ASME	American Society of Mechanical Engineers
ASTM	American Society for Testing and Materials
AWES	Association of West European Shipbuilders
BP	British Petroleum
CAAM	Centre administratif des affaires maritime
CAP	condition assessment program
CDS	construction differential subsidies
CFR	Code of Federal Regulations
CGT	compensated gross ton
CIALA	Centro de informacion del acuerdo latinamerico
DB	double bottom
DNV	Det Norske Veritas

DOT	U.S. Department of Transportation
DS	double sides
DWT	deadweight ton
EIA	Energy Information Administration, U.S. Department of Energy
FSU	former Soviet Union
GRT	gross registered tonnage
GT	gross tons
H&M	hull and machinery
HBL	hydrostatically balanced loading
HTS	high tensile steel
IACS	International Association of Classification Societies
ICLL	International Convention on Load Lines
IMO	International Maritime Organization
INTERTANKO	International Association of Independent Tanker Owners
ISGOTT	International Safety Guide for Oil Tankers and Terminals
ISM	International Safety Management (code)
ISO	International Standards Organization
ISOPE	International Society of Offshore and Polar Engineers
ITB	integrated tank barge
JAMRI	Japanese Maritime Research Institute
LOOP	Louisiana Offshore Oil Port (a deepwater offshore port)
LTBP	London Tanker Brokers' Panel
M&R	maintenance and repairs
MARAD	U.S. Maritime Administration
MARIENV '95	International Conference on Technologies for Marine Environment Preservation, Tokyo, Japan, September 24–29, 1995
MARPOL 73/78	International Convention for the Prevention of Pollution from Ships, adopted in 1973 and amended in 1978
MBD	million barrels per day
MEPC	Marine Environmental Protection Committee of the International Maritime Organization
MEPC30	30th session of the Marine Environmental Protection Committee
MIT	Massachusetts Institute of Technology
MMS	Minerals Management Service, U.S. Department of the Interior
MOU	memorandum of understanding

MPA	Marine Preservation Association
MSI	Marine Strategies International
NASA	National Aeronautics and Space Administration
NAVSEA	Naval Sea Systems Command
NOAA	National Oceanic and Atmospheric Administration
NRC	National Research Council
OBO	oil-bulk-ore (vessel)
OCIMF	Oil Companies International Marine Forum
ODS	operating differential subsidies
OPA 90	Oil Pollution Act of 1990 (P.L. 101-380)
OSG	Overseas Shipholding Group, Inc.
P&I	protection and indemnity
PIRA	PIRA Energy Group
PL	protectively located
RFR	required freight rate
SBT	segregated ballast tank designed for ballast only
SIRE	Ship Inspection REport
SNAME	Society of Naval Architects and Marine Engineers
SNPRM	supplemental notice of proposed rule making
SOLAS	International Convention on Safety of Life at Sea
STCW	Convention for Standards for Training, Certification, and Watchkeeping
TAPS	Trans-Alaska Pipeline System
TSCF	Tanker Structure Cooperative Forum
ULCC	ultralarge crude carrier
USACE	U.S. Army Corps of Engineers
USCG	U.S. Coast Guard
VLCC	very large crude carrier
WS	Worldscale

GLOSSARY

Compensated gross ton. Term defining the capability of a shipyard that reflects the complexity of construction of a vessel.

Deadweight tonnage. Measure of the weight of cargo (plus water, fuel, and stores) that a vessel can carry.

Gross tonnage. Measure of a vessel's volume determined according to international convention.

Hydrostatically balanced loading. Means whereby the level of cargo (e.g., crude oil) is limited to ensure that the hydrostatic pressure at the tank (and ship) bottom is lower than the external sea pressure. Thus, if the tank is breached, seawater will flow in rather than oil flowing out.

International Maritime Organization. United Nations agency responsible for maritime safety and environmental protection of the seas.

Lightering. Process of transferring cargo at sea from one vessel to another.

Oil-bulk-ore vessel. Vessel designed for alternate carriage of oil, bulk cargoes, or ore.

Required freight rate. The rate required to cover tanker operating expenses and realize a desired return on capital.

Segregated ballast tank. Tank designed for ballast only.

Ultralarge crude carrier. Vessel of more than 400,000 DWT.

Very large crude carrier. Vessel of between about 150,000 and 300,000 DWT.